W9-ASF-207

RESTRUCTURING SCIENCE EDUCATION

The Importance of Theories and Their Development

RESTRUCTURING SCIENCE EDUCATION

The Importance of Theories and Their Development

RICHARD A. DUSCHL

Teachers College, Columbia University
New York • London

Published by Teachers College Press, 1234 Amsterdam Avenue
New York, NY 10027

Library of Congress Cataloging-in-Publication Data

Duschl, Richard A. (Richard Alan), 1951–
Restructuring science education : the importance of theories and their development / Richard A. Duschl.
p. cm.
Includes bibliographical references (p. 139)
ISBN 0-8077-3006-8 (alk. paper).
ISBN 0-8077-3005-X (pbk. : alk. paper)
1. Science—Study and teaching. 2. Curriculum planning.
I. Title.
Q181.D78 1990
507.1—dc20 89–20414
CIP

ISBN 0-8077-3005-X (pbk.)
0-8077-3006-8

Printed on acid-free paper
Manufactured in the United States of America

97 96 95 94 93 92 91 90 8 7 6 5 4 3 2 1

For Marylou,

who provided the support and a working climate that made the preparation and writing of this book possible

Contents

Foreword

- In the interest of fair play and equal time, creation science should be part of school science programs.
- Astrologers should provide advice to heads of state concerning their travel, meetings, and policy decisions.

What's wrong with these statements? How can intelligent and well meaning people make such statements? There is more than one important point underlying such issues and questions. One point is that all individuals need explanations for their world, and religion and astrology both provide explanations. Science also provides explanations about the world. As people develop explanations for their world, they may not necessarily differentiate between the sources or development of those explanations. They only know that they have explanations that make their world understandable. A second point is that these explanations may be in conflict, as is the case in the ongoing science vs. creationism controversy; or, explanations may develop in the absence of any other clear explanations, as is the case when people rely on daily examination of astrologers' charts.

One must also acknowledge the fact that many individuals have deep emotional commitments to their explanations of the world and that confrontation with another, conflicting, explanation engages more emotion than rational analysis. Dissonance theory in psychology informs us that individuals cannot believe both A and B when those are in conflict. Our tendency is to hold tenaciously to the original idea; generally we seek evidence supporting it, as opposed to evidence refuting or modifying our current conceptions.

What does this discussion have to do with Richard Duschl's timely book *Restructuring Science Education: The Importance of Theories and Their Development*? While the emotional and psychological dispositions of individuals must be acknowledged, it is also the case that scientific information and rational understanding will contribute to the resolution of conflicts such as those identified in the first paragraph. Some important goals of science education are to clarify what science is and what it is not; what science can do and what it cannot do; what science can explain and what it cannot explain; what rules and procedures scientists obey; what happens when individuals do not obey the rules; and finally, what individuals and society ought to expect from science and which expectations are inappropriate. If these are among the important aims of science education, we can ask whether science teachers understand the goals and if these goals are addressed in contemporary programs. Unfortunately, the answer to both these questions is *no*. Science teachers have considerable facts and information *of* science. And that is what they teach. Science teachers, like most citizens, have little understanding *about* science. So, they teach very little *about* science. How can this change? You have one answer in your hands. This book provides an excellent introduction to ideas such as the role of observation in science, the processes of science, the growth of knowledge, the place and displacement of theories in science, and the initial, but often neglected, place of history in science education.

In this book, Dr. Duschl also makes the connection between the development of theories and the development of students' understanding of the world. He demonstrates the need for conceptual change in teaching, and the parallel instructional strategies needed to facilitate conceptual change in the work and development of science and scientists. Science teaching that engages the learner, allows exploration and challenge of ideas, and provides new explanations is congruent with the nature of science itself and good science teaching. Developing a sense of explanatory richness and broader applicability of scientific theories is what science and science teaching should be about.

Restructuring Science Education first reviews the last 30 years of science education, and in the process clarifies the philosophical orientation that is most prevalent in today's school science programs. In the same 30 years, there have been significant developments in the history and philosophy of science. Dr. Duschl shows how those developments are not evident in contemporary science education. His book then addresses the central role of theories and uses theory as the conceptual framework to incorporate other aspects of science, such as the development and justification of new knowledge. The conclusion suggested by Dr. Duschl is that science teachers should consider the structure and development of theories as an organizing framework for curriculum and instruction. The book develops this recommendation across eight chapters and does so in a way that really helps the reader learn more about both the history and nature of science, and about different and innovative approaches to instruction.

In this era of education reform, scientific literacy has become a theme for today's science education. Scientific literacy certainly includes developing a better understanding of the history and nature of science and technology. Achieving this goal could help both scientists and nonscientists to resolve social and public policy issues like those described in my opening paragraph and those confronting individuals every day. In our age, an enlightened citizenry needs to understand the nature of science as integral to our society. Thus, better science teaching is essential.

Richard Duschl's *Restructuring Science Education* joins Jim Robinson's *The Nature of Science and Science Teaching* (1968) and Michael Martin's *Concepts of Science Education: A Philosophical View* (1972) as my recommendations for tools for including philosophical views in school science programs. Historically, science education has recognized the goals of the nature of science, but science programs and science teachers have done little to develop those goals. This book is a contemporary opportunity to incorporate a vital goal of science education.

RODGER W. BYBEE
ASSOCIATE DIRECTOR
BIOLOGICAL SCIENCES CURRICULUM STUDY
COLORADO SPRINGS, COLORADO

Preface

This book is intended to serve as a companion volume to science education methods texts. Its contents fill a void found in many existing texts. In the opening two chapters, science education is placed within a recent historical context. The activities after the formation of the National Science Foundation, namely, those in the history and philosophy of science and in cognitive psychology, are reviewed and put into a broader context with other relevant activities concerning the study of science. Following a review of the developments in science education, history and philosophy of science, and cognitive science, procedural guidelines for joining these disciplines are outlined. The next four chapters outline examples of how the procedural guidelines for testing theories and for characterizing the growth and restructuring of scientific theories might be used in the design and implementation of curriculum units on the growth of scientific theories. The last two chapters provide specific examples from physics, chemistry, biology, and geology for using the frameworks.

Hence, the book might also be used as a companion volume in a curriculum theory course. Here the focus is decidedly one that addresses the curriculum question, "What to teach?" from an epistemological perspective. The thesis presented here is that an analysis of the growth of scientific theories can guide decisions about what the most important content is. This is one solution to the depth vs. breadth curriculum problem.

The teaching and learning of science have risen in stature over the last 30 to 40 years to occupy a place of vital importance in education and in society. Although much has changed in the curriculum approaches and the theories of learning advocated for the successful teaching of science, the classroom teacher remains the centerpiece of the enterprise.

It is a fundamental assumption of this book that the classroom teacher ought to be a decision maker. This point is repeated over and over throughout the text. Another fundamental assumption is that the processes used in the teaching and learning of science are maximized when the procedures for designing and implementing instructional content are congruent with the procedures students are asked to employ in learning the content. The application of epistemological (knowledge growth) models to the design and implementation of science instruction meets this need.

Science teachers are fortunate in that the subject matter they teach is the result of the same type of activities we ask our students to employ. Science is a disciplinary activity that seeks to understand the how and why of natural phenomena. It does this by developing explanations we call scientific theories. The study of the growth of scientific knowledge by philosophers, historians, and sociologists of science has revealed that scientific theories are at the core of the process. Similarly, through the study of cognitive processes we have learned that an individual's conceptual frameworks serve as personal theories that guide the learning process.

The commonplace between the subject matter of science and the learner is theories. The thesis of this book is that procedural guidelines that philosophers of science have proposed for explaining the structure and restructuring of scientific theories can be employed as tools by teachers. One use is as an aid in planning and making decisions about the selection and sequence of science instruction. We learn, for example, from the study of scientific theories that all knowledge claims are not equally important. Thus, growth-of-knowledge frameworks help us to decide what truly is the most important content. Another use is as a learning aid for students. When teachers and students begin to employ similar procedural steps and talk the same language in the processes of teaching and learning science, the learning environment improves. Employing growth-of-knowledge procedures from the history and philosophy of science makes this possible.

Acknowledgments

It was 10 years ago that I learned that the perspective of the growth of scientific knowledge that I had developed from my training as a science teacher was not only inaccurate but drastically off target. Thank you, Philip Ehrlich. Since that time I have made myself a case study of one and focused my attention on ways to better integrate concepts from the philosophy of science with science education. While initially it was a lonely struggle, I was fortunate to have many beneficial relationships with scholars along the way. A special thanks to Emmett Wright, who agreed to work with a wayward doctoral student.

I owe a great deal to the encounters I had with Stephen Brush, Fred Suppe, and Dudley Shapere during my doctoral studies at the University of Maryland. Little did I realize then, as a neophyte, how fortunate I was to study with these individuals. I thank them for accepting a science educator into the fold and for enabling me to maintain a scholarly dialogue over the years.

Upon my acceptance of an assistant professorship at the University of Houston, Dudley Shapere recommended I look up Richard Grandy at Rice University. A special thank you to Dudley because Dick, through our hours and hours of critical discussions together, helped to mold me into a better writer and a thinker of original ideas. I am also indebted to Howard Jones, Mark Ginsburg, and Richard Hamilton, who as colleagues at Houston introduced me to new ways of thinking about science education, education, and learning, respectively. Each read more than his fair share of draft manuscripts that led to the development of the ideas and arguments in these pages.

Also, a thank you to Teachers College Press editors Sarah Biondello, Susan Liddicoat, and Nina George, for making the preparation of my first book a pleasurable experience. Finally, I would be remiss if I did not acknowledge the important contributions of my parents, and daughters Cary, Gretchen, and Rebecca, who over the years have provided me the much needed diversions of play and relaxation, and of my friends Paige, Brin, and Hersh for being such good listeners.

RESTRUCTURING SCIENCE EDUCATION

The Importance of Theories and Their Development

Chapter 1

Bases of Science Education

Ask a science teacher what he or she could use more of, and a common answer will be time. It is a well understood maxim of classroom teachers that effective and meaningful science instruction that seeks to create a hands-on, inquiry-oriented learning environment requires a great deal of time. So too do instructional units that seek to develop higher level thinking skills. It is precisely for this reason that advocates of new approaches in science education are calling for an abandonment of the survey type science courses, which attempt to teach a little information about a lot of topics, in our secondary schools.

The call is to replace the survey courses with courses that examine select themes of science in depth. What is gained? A richer understanding by students of how the concepts of science are interwoven into a fabric of scientific knowledge becomes possible. The interconnection of concepts stresses students' understanding of the relationships among scientific concepts. Another positive outcome is that a comprehensive introduction to the use of higher cognitive thinking skills becomes possible. Like any other activity, thinking needs to be broken down into a set of procedures students can follow; then these procedures need to be practiced. Limiting a science curriculum to a small number of themes allows time for the in-depth analysis, synthesis, and evaluation that constitute high level thinking skills.

What is gained for the teacher? For teachers who are skilled decision makers in the selection of the most important science content and who are also skilled in sequencing instructional tasks based on the needs of learners, time in the curriculum is created.

TEACHERS AS DECISION MAKERS

Teaching occurs within a very complex environment. It takes a skilled individual to establish an effective learning environment given the diversity of students and the variety of educational tasks he or she must face. Right or wrong, appropriate or not, a teacher makes numerous decisions on a daily basis concerning the design, delivery, and evaluation of instruction. An effective decision maker considers the learner, the learning environment, and the nature of the subject matter. The issue is a teacher's ability to make informed judgments that eventually lead to meaningful learning on the part of students. A central component of this decision-making process is a teacher's subject matter knowledge—knowledge that embraces not merely the facts, principles, and definitions of science but also an understanding of the structures of the science discipline. Knowledge of the structures of a discipline—the guiding conceptions, relevant research questions, and procedures used to design, carry out, and evaluate scientific investigations—empowers a teacher with a special set of decision-making skills. The teacher can then determine the proper emphasis to place on knowledge claims and select the procedures, both for teaching and learning, that ought to be employed to achieve the goal of meaningful learning by students.

When a teacher is capable of selecting, sequencing, and implementing science lessons so that the curriculum reflects a prioritization of content and knowledge objectives, time is created for meaningful learning, for teaching the procedures of learning, and for teaching higher level thinking skills. The payoff is that students become well informed about what scientific knowledge was or still is most important, what research questions or problems were or are most relevant, and what criteria are considered appropriate in evaluating scientific knowledge; at the same time, students employ sound reasoning patterns in the analysis, synthesis, and application of scientific knowledge.

To achieve such educational goals, the teacher needs to acquire a special type of knowledge base. In addition to knowledge of schools, of learners, and of teaching strategies, teachers also need a special knowledge base of the structure of the subject. This type of knowledge is the basis for subject matter prioritization decisions and learning procedures unique to science. One objective of this book is to outline instructional frameworks teachers can use in making decisions concerning the prioritization of science content; another is to introduce guidelines teachers can use to determine the sequence of instructional tasks. Thus, this book is about proactive and interactive curriculum design decision making.

The commitment is to the teacher, to the realization that

teachers who are capable of modifying and adapting instruction to meet their needs, and, more important, the needs of their students, are the most effective teachers. What distinguishes this book from other science "how-to books" are (1) the source of the frameworks used to help teachers modify and adapt instruction, and (2) the basis used to think about the teaching of science. The source of the frameworks is the recent work of historians and philosophers of science—scholars, not unlike science teachers, who attempt to understand and communicate the essence of the scientific enterprise. Their understandings offer enlightening new perspectives on what ought to be considered the most important content. The approach, derived from the source, is emphasis on a commitment to the development of scientific knowledge—a development that is not dissimilar from the mechanisms of change that occur within individual learners. This will show how historians' and philosophers' considerations of the growth of scientific knowledge can also be used as a basis for instructional design and as a source for the selection of procedures that learners can use to evaluate the legitimacy of scientific knowledge claims.

It is a goal of this book to fill a void that exists in other secondary science education methods texts. That void is the lack of an in-depth presentation of the recent history of science education and the scant coverage given to topics of epistemology, that is, the study of the structure and nature of knowledge. From the examination of the history of science education, teachers will be given a new perspective to guide science practice. The study of the structure of knowledge will provide a set of decision-making tools that will enable you to be architects of scientific knowledge. As architects, you will acquire knowledge of the materials and methods needed to select, design, and implement instructional tasks that are meaningful and develop critical thinking skills.

The remainder of this chapter will outline, in broad terms, the theme of the rest of the book. This overview of the critical elements of the science teaching and planning approach I am advocating will enable readers to understand how the remaining chapters are related. In those chapters, details will be added to complete the task of bringing theory into practice.

CRISES IN SCIENCE EDUCATION: FAMILIAR TERRITORY

The importance of a sound science education has, over the past 35 years, been discussed, debated, and recognized as an essential element for the citizens of today and tomorrow. For the second time since the conclusion of World War II, a crisis in science education has been declared. In 1982, John Slaughter, then Director of the National Science Foundation, talked of an ever-increasing gap between the scientific elite and the scientific illiterate. He pointed out that citizens found themselves in a time when the new knowledge generated by scientists was quickly outpacing the public's abilities to keep up. The resulting challenge to teachers and curriculum writers is how to close this chasm between what is known by experts but not by teachers, students, and citizens. At the very center of this controversy are the changes in procedures and data sources scientists employ in their pursuits of knowledge—changes that in themselves are, at times, more astonishing than the new knowledge being proposed. Examples are the procedures doctors now employ to analyze and diagnose internal medical problems—sonograms, CAT scans, optical fibers; and the ways that computers and microchips have changed the landscape of our lives.

The implication of these rapid changes in knowledge for science teachers is twofold. First, we must recognize that as teachers of science we will be faced with more instances when it will be necessary to modify the science curriculum being taught to students. Hence, the problem of determining the most important content will not only persist, but gain importance in the years to come. As advances are made in our fundamental understanding of the world through the development and application of investigative processes—which in themselves represent significant advances—we science teachers will find ourselves burdened with the important task of keeping up. Otherwise we will not be able to meet our commitment to sound and informed decision making.

This rapid acceleration of scientific knowledge and technology also creates the problem of how to make our instruction reflect the rational nature of science in the face of change. In other words, we have arrived at a time in history in which the knowledge taught to students in high school or postsecondary schools can, on occasion, be quite different from the knowledge taught to the same students during the elementary or middle school grades. The differences are not in the degree of complexity, but in fundamental beliefs; different explanations are provided and different guiding concepts determine the methods and standards used in investigating problems and inquiring into research questions. Such changes involve new ideas replacing older ideas and are common in science.

Consequently, a second challenge for us as science teachers educating citizens of the twenty-first century is how to design science instruction that reinforces rather than ignores what we intuitively know will happen to scientific knowledge—that it will change. Essentially, then, we are faced with the task of providing instruction that makes such change a goal of science education. In short, I am speaking of the need to explore the nature of science and scientific inquiry in science classrooms. Such an exploration will enable teachers to apply new models of teaching science and to improve the efficiency of learning in their classrooms.

HISTORY, PHILOSOPHY, AND A NEW FACE OF SCIENCE

As educators we are fortunate that others have chosen to explore this changing, evolving aspect of science and have generated a wide array of useful guidelines for teachers to use in the selection, sequencing, and evaluation of instructional tasks. One factor that distinguishes attempts to improve science education today from attempts made in the 1950s and 1960s is a richer understanding of the processes associated with the growth of knowledge. It is richer because we have access to the work of historians and philosophers of science, which has helped to develop the position that the growth of knowledge in science is best understood as a series of changes in scientists' fundamental explanations of how and why things work. Today we recognize these fundamental explanations as scientific theories, and the changes as conceptual changes in the composition of scientific theories. By analyzing the work of historians, philosophers of science discovered three things about the nature of science (Shapere, 1984).

1. The standards used to assess the adequacy of scientific theories and explanations can change from one generation of scientists to another.
2. The standards used to judge theories at one time are not better or more correct than standards used at another time.
3. The standards used to assess scientific explanations are closely linked to the then-current beliefs of the scientific community.

Thus, contemporary studies on the nature of scientific inquiry focus on how theories of science come to be, how they proceed through stages of development, and how they are replaced.

The upshot of all this for science educators is that it is now possible to speak about science knowledge in terms of a logic of development. The importance of this position will be addressed more fully later. Suffice it to say here that there are guidelines used in the process of acquiring scientific knowledge that are quite similar to guidelines employed in the process of learning. It will be seen that the view of science that arises from the work of historians and philosophers of science has generated concepts that resemble the newer views cognitive scientists propose for how children learn science. Simply stated, both the processes of learning and the growth of knowledge in the field of science involve mechanisms in which new ideas replace old ideas. This marriage of psychological principles and epistemological principles forms a foundation for the recommendations for changes in science education methods made in this book.

Prior to the 1950s, the nature of science was understood through the analysis of only the most highly developed of the sciences—physics and chemistry—and the examination of fully developed theories of science. The examination of such explanations (Newton's laws of motion, for example) established a view of science that made understanding the testing of knowledge claims much more important than understanding how knowledge was discovered. It was considered unimportant to try to understand how ideas were first conceived and how they evolved over time with the inevitable discovery of new evidence.

Consideration of other disciplines or domains of science, however, has indicated that fundamentally different sets of criteria are employed by scientists within different disciplines. The goal is always the same—an understanding of the nature of the world—but different guiding conceptions give rise to different methods of obtaining explanations and solving problems. Basically, scientific inquiry is not as neat and clean as science educators often propose. Rather it is in a continual state of flux in which theories, methods, research questions, and so forth compete for supremacy. But this is not to say it is totally devoid of standards or frameworks for judging competing ideas and choosing among them. It merely establishes that a great deal more needs to be taken into consideration and a great deal more ought to be embodied in our science classes.

SCIENTIFIC THEORIES

It is a common mistake among laypeople to equate scientific theory with scientific fact. Indeed, many individuals will argue that facts are more important than theory. Nothing could be further from the truth concerning how science operates. Scientific theories are the most important element of scientific knowledge and play a central, vital role in the growth of scientific knowledge. However, interest in understanding the facts, labels, and quantities involved in science often takes precedence in our classrooms. Although measurements, descriptions, drawings, and photographs of the physical and biological world constitute the database of science, such knowledge represents a commonsense notion of science. Facts in themselves do not provide understanding. It is the relationships among facts that develop into our representations of understanding. Scientific knowledge, as distinguished from commonsense knowledge, does not deal with answers to "what" questions. Rather, scientific knowledge involves attempts to understand and explain why something occurs or exists the way it does, or how something occurs. It is from this synthesis of facts—drawing inferences from what is known—that scientific theories arise and are altered.

Understanding how theories develop and affect other aspects of science enables science teachers to draw upon a very powerful set of guidelines in their curriculum design decision making. Contemporary analyses of science suggest that scientific theories progress through stages. In other words, they have a developmental history, which can be evaluated and examined. Very significant to us as science teachers are the realizations that

1. All scientific explanations are not equal; some ideas are more important than others.
2. A scientific explanation can rise and fall from grace in the scientific community.
3. Criteria exist for judging or evaluating scientific explanations.
4. A description of the rational evolution of scientific explanations is possible.

The analyses made by historians, philosophers, and sociologists of science of the changes that occur in scientific knowledge make it quite clear that an understanding of the growth of scientific knowledge is best obtained through an understanding of the development of scientific theories. Thus, a better understanding of scientific reasoning and of the factors that contribute to the growth of scientific knowledge can be acquired through judging the structure of scientific theories. It is possible to judge scientific theories, that is, to evaluate and compare competing ideas. As will be explained in Chapter 5, there are at least four fundamental criteria that can be employed to judge a scientific theory (Root-Bernstein, 1984). These criteria address the logical, empirical, historical, and sociological aspects of scientific explanations. Taken together the four criteria outline a means to assess the strengths and limitations of scientific knowledge claims.

TWO FACES OF SCIENCE: TWO GOALS OF SCIENCE EDUCATION

The activity of science—as it is represented by the growth of knowledge—involves two equally important sets of processes. First there are the processes associated with the generation of scientific knowledge claims. Also called the "context of discovery," these processes address the developmental characteristics of scientific knowledge. The context of discovery, then, involves the origin and evolution of ideas.

An example would be the events that resurrected the idea of continental drift in the 1950s. Dismissed in 1926 by geologists as an incorrect model, the revival of the theory of continents moving with respect to one another involves physicists more than geologists. The key to the rediscovery was physicists' discovery of remanent magnetism in rocks (the aligning of iron minerals in the prevailing direction of the earth's magnetic field at the time the rocks were formed). The historical and sociological criteria of this episode in science contribute significantly to defining the context in which the restructuring of knowledge actually occurs.

The second set of processes are the more familiar ones for the justification of scientific knowledge. These processes in the growth of scientific knowledge are associated with the testing of scientific-knowledge claims and come under the heading of the "context of justification." Gathering and establishing the validity and reliability of scientific evidence are addressed by the context of justification. It involves the direct application of logical and empirical criteria science uses to legitimize its knowledge claims. Continuing with the example of continental drift, the testing of knowledge claims would be the investigations that established that the rocks of the ocean floors

1. Are found in wide bands with magnetic reversals.
2. When of the same age, occur equally distant from both sides of the mid-ocean ridge and have the same magnetic directions.
3. Are much younger than rocks on the continents.
4. Are found to decrease in temperature as they move away from the mid-ocean ridge.

In each case, observable evidence was obtained to establish the claim that the earth is composed of rigid plates, which drift apart from spreading zones.

Our attempts as teachers and as students to understand the nature of science are complicated by the fact that the processes, just described, of discovering and justifying knowledge occur within a context of known and accepted knowledge. In other words, science has a set of standards that must be learned, and as explained in the previous section, the history of science has shown that these standards change. Thus science has two faces, or two profiles to the same face. On one side we find the products of science: the facts, principles, laws, and theories that make up the knowledge base and set the standards of science. On the other side we find the processes of science: the methods employed in the collection, analysis, synthesis, and evaluation of evidence. It is important to stress, because it is often neglected, that the processes of science include both manipulative and cognitive processes.

This two-faced nature of science has dominated science education practice during the twentieth century. From the product characterization of science, we obtain the concept approach to the design of science instruction. Focusing on the investigative procedures of science, we obtain the inquiry and process approach to science. There is a problem, however: The processes of science as portrayed in the majority of elementary and secondary science textbooks have focused almost exclusively on activities associated with the context of testing. What results, then, is an incomplete representation of science. "Epistemologically flat" (see Kilborn, 1980) is one way to put it, with flatness referring to the lack of knowledge concerning how we have come to be at this particular juncture in our understanding of the way the world works. What is missing is the chain of reasoning that has brought us to this point of understanding.

It is the testing of knowledge activities, then, that presently dominates contemporary science education practice. Scientific-knowledge claims—facts, hypotheses, principles, theories—are learned on the basis of their contribution to the established or final form models of knowledge. How such knowledge came to exist is treated as a non-issue in most science textbooks and curricula. A science curriculum that focuses solely on the prevailing knowledge claims or views of science merely teaches scientific knowledge. To paint a complete picture of science, a science curriculum should address not only what is known by science, but also how science has come to arrive at such knowledge. To teach what is known in science requires curriculum objectives seeking scientific knowledge. To teach how the scientific enterprise has arrived at its knowledge claims requires curriculum objectives seeking knowledge about science.

The distinction may seem to be trivial, but it is not. Knowledge about science, as opposed to scientific knowledge, is knowledge of both why science believes what it does and how science has come to think that way. In a knowledge-about-science curriculum the interactions among science, technology, and society are much more relevant and thus are more easily appreciated. It attempts to stress the most important content, to introduce the guiding conceptions of science, and to establish in learners the ability to evaluate the legitimacy of knowledge claims.

The distinction between scientific knowledge as a curriculum objective and knowledge about science as a curriculum objective can be explained in this way. We can call scientific-knowledge teaching *S1* and knowledge-about-science teaching *S2*. *S1* is a portrayal of science that reports observations and abstract theories that account for them. *Reports* is the key word in the previous sentence because how science is reported and how it is actually carried out are quite different. From the reports, we obtain a view of science that is logical and always on the mark. It is a dialogue of the successes of science, and it presents the final form of the present view of the world.

S2, on the other hand, is a reporting of how discoveries are made. It examines the various false starts, modification of concepts, provisional and finished claims of knowledge, and reasoning employed in the pursuit of understanding. It is very important to realize that both aspects of science teaching—the final form elements and the stage development elements—are critical. But, while *S2* must draw upon *S1* to fulfill its objectives, *S1* need not include information from *S2*. Therein lies the problem—a problem of exclusion. It is certainly necessary to include instruction on the products of science, but that instruction is insufficient for teaching students about the rational evolution of scientific knowledge: "The question of why science today believes the peculiar things it does about the universe, and why it is willing to consider the alternatives it does, requires attention to the question of how science has come to think in those ways" (Shapere, 1984, p. 190).

By choosing to neglect the fundamental concepts of scientific change and not to examine in our classes the warrants or reasons scientists use to change scientific methods, beliefs, processes, and so forth, we are running the risk of developing students who do not acknowledge scientists' views as rational and as the end product of a process in which changes are both natural and expected. If a person cannot understand, or at least appreciate, what counts as scientific evidence, then what hope is there that that same person will understand the conclusions drawn from that evidence. We need not look long and hard to see that this is occurring today. Consider the support for and the notoriety of arguments for the teaching of creation science in the United States. Consider, too, the uninformed reactions individuals have had about people infected with the AIDS virus.

As the processes of science used in gathering and evaluating scientific evidence become more sophisticated, the need to establish a curriculum that examines the chain of reasoning that has brought us to this point gains in importance. Each school year, a minimum of one instructional unit—four to six weeks of instruction—should focus on the context of discovery. The best theme or context for such an instructional unit is one of the theories scientists embrace as a central component of their discipline.

THE NATURE OF THE LEARNER IN LEARNING SCIENCE

The tragedy of teaching science as absolute truth or with curriculum objectives that seek to empower students with the contemporary final form knowledge of science is that the potential to achieve something very special with students is lost. For the teacher who is capable of making curriculum design decisions informed by considerations for the context-of-discovery elements of science, the potential to provide students with a meaningful science learning experience is greatly enhanced. A strong theme within this book is that learning as it occurs within individuals is guided by the same basic sets of principles that guide the growth of knowledge in science. Such a claim was the foundation of Piaget's pioneering work in understanding the cognitive development of children, and it continues to guide much of how educators and psychologists understand and interpret learning. There is much to be gained by developing a strong link between psychology and epistemology—the two branches of study that investigate the growth of knowledge in individuals and the growth of knowledge in disciplines, respectively. It seems intuitively correct to suggest that the reasons for changes in scientists' views may also help evoke changes in children's views: "An understanding of the critical historical experiments which led to scientists' changing their ideas may also have implications as to how children's views might be changed" (Osborne & Freyberg, 1985, p. 108).

What is missing in many science teacher education programs is a sound introduction to how the structure of knowledge, the procedures of logic, the methods of empirical analysis, and the criteria used to judge scientific theories can be linked with psychological principles of learning and used in making decisions about the design of instruction. Recent research on students' understandings of basic science concepts has found that many students harbor very wrong ideas about how the world works. These views are also referred to as misconceptions or naive theories, and have the characteristic of being significantly different from scientists' views of the world. Learning in science and cognitive development in general are conceived as processes in which old ideas, concepts, and meanings are replaced by new ones. Susan Carey (1985a), a leading researcher in how children learn science, proposes that cognitive development is a process involving conceptual changes. The challenge for science teachers is how to design instructional strategies that will promote the evolution of students' naive views into the more sophisticated scientists' views. This type of teaching has been labeled "conceptual change teaching" and offers a strong alternative to existing models of science teaching (Anderson & Smith, 1986; Novak & Gowin, 1984; Osborne & Freyberg, 1985).

What distinguishes the conceptual-change-teaching model from other models of teaching science (discovery teaching, processes-of-science teaching) is the recognition that all learning begins with and is subsequently influenced by the prior knowledge of the student. Conceptual-change teaching will be discussed in more detail in Chapter 6. Basically it is a view of science teaching that argues that cognitive development in learning science involves a series of changes in the knowledge frameworks children employ (Posner, Strike, Hewson, & Gertzog, 1982). To bring about the change, however, learners must first become dissatisfied with their existing point of view. Next the new idea must be internally consistent—make sense to the children. Thus, the more sophisticated scientific point of view must be seen as a more plausible and more intelligent view than their own. Finally the new idea must be demonstrated in new situations to establish why it is a stronger explanation or concept.

This model advocates that children's views of science concepts be evaluated not only against scientists' views but also against the views of their classmates. A hallmark of this teaching approach is the active participation and dialogue that occur between teacher and students and among students. By necessity, the dialogue must identify initial concepts, challenge those that are inadequate, introduce the more sophisticated concept, and then demonstrate how the new concept contributes to a fuller and richer understanding of the natural world. The challenge to teachers of science is that special teaching strategies need to be used that reflect a comprehensive understanding of the structure of science. It follows, then, that teachers need new guidelines to help make decisions concerning the selection and implementation of instructional tasks. Good teaching is always preceded by good planning.

OVERVIEW OF REMAINING CHAPTERS

For the past 30 years, science education has emphasized a curriculum focusing on science for future scientists. The focus is based on a philosophy of science that emphasizes the justification of knowledge. From this focus have come two significant and dominant science teaching strategies: (1) the process approach, which emphasizes generic skills and techniques science uses to collect, manipulate, and interpret data, and (2) the inquiry approach, which emphasizes the role of hands-on activities and investigations and the student as an active learner. Although other approaches to science education exist, these two dominate much of the recent curriculum writing being directed by state education agencies. There is a concerted effort, however, to generate new models of teaching and learning science. One characteristic of the new models is the recognition that principles of cognitive science and principles of epistemology have much in common. Integrating these principles into the decision-making guidelines teachers employ when planning instruction is an important aspect of science teacher education programs.

The chapters that follow demonstrate how the history and philosophy of science can be applied by teachers to the effective planning and teaching of science. Chapter 2 will examine the first science education crisis of the 1950s and the teaching and curriculum themes of the National Science Foundation projects. Chapter 3 explores the second crisis in science education from the early 1980s, which occurred about 25 years after the first. It outlines the shift in science education emphasis from a "science for scientists" focus to a "science for all" focus. Chapter 3 also outlines the important developments in the history of science that altered fundamental tenets in the philosophy of science. Of particular interest are the changes that occurred in our conceptions of scientific theories and of the role that theories play in the growth of scientific knowledge. Chapter 4 addresses the central role of theories in the growth of scientific knowledge and provides a fuller distinction between the two fundamental contexts of science, the context of discovery and the context of justification. Chapter 5 extends the discussion of theories. Here we examine the changing nature of scientific evidence and how over time we have learned how to learn. The implication is that the practice of science has evolved to the point where today scientists depend less and less on the direct use of their senses for the implementation of basic scientific processes (such as, observing and measuring). As a result, scientific data may become more inaccessible to students in precollege science classrooms.

Chapter 6 synthesizes the major themes of earlier chapters for the purpose of advancing the argument that teachers should consider the structure and development of theories in their decision making concerning the design, implementation, and evaluation of instructional tasks. Here theory will be put into practice. In Chapters 7 and 8 specific examples of the application of these guidelines are outlined for each of the principal secondary level science disciplines—chemistry, physics, earth science, and life science.

Chapter 2

A Recent Retrospective of Science Education

In 1982, a convocation was held at the National Academy of Sciences and Engineering (1982). It was not a gathering called for the purpose of celebrating an event or anniversary. It was a political event, which sought to focus the nation's attention on the state of affairs in precollege science and mathematics education. At that two-day gathering, politicians, scientists, captains of business and industry, and educators gave speeches and made arguments that the United States was again facing a crisis in science and mathematics education. The evidence was quite strong that we were losing our competitive edge in the world's science and technology marketplace. For the second time since the conclusion of World War II national attention had been focused on the state of affairs in science education.

What is odd is that the crisis in 1982 came a mere 25 years after a major national effort to upgrade the quality of science curricula and instruction. This effort saw the development of new curricula and new support materials at a funding level involving hundreds of millions of dollars. Thus it was peculiar that the opening speaker of the 1982 convocation, Frank Press, President of the Academy, identified the starting point of the erosion of mathematics and science education as the year 1962. In 1962 the National Science Foundation (NSF) was five years into the largest funding effort ever undertaken by any nation to improve science and mathematics instruction in schools.

It is ironic that the downfall of science and mathematics education would be identified as starting in the midst of the 1960s curriculum reform movement. What happened at that time to prompt such a comment? What lessons might we learn from others' efforts to improve science education? A historical analysis of events in the science curriculum development efforts leading up to 1962 will help shed light on the situation. This chapter will examine both the issues and the developments surrounding the design and implementation of the early NSF curriculum projects. The intent is to provide the reader with a background for understanding the issues that confronted science educators in the 1950s, issues that we seem to continually face. The issues can be broadly categorized into those addressing the reeducation of teachers and those addressing the redesign of curriculum.

THE POSTWAR YEARS

The United States and the Allied forces emerged from World War II not only with a military victory over the Nazis and the Japanese but also with their scientific and technological superiority intact. As much as anything else, the scientific know-how and technological wizardry of the United States contributed to the winning of the war. It was because of the impressive technological successes of World War II that NSF was created in 1950. These successes (radar, sonar, nuclear energy, jet airplanes, artificial rubber, to name but a few) made quite evident the important role technology would play in establishing the political, economic, and social health and strength of the United States in the years ahead. The National Science Foundation was charged with guaranteeing that our nation's potential in science research and science education would be exemplary.

Initially, the role of NSF in science education was limited to in-service education of college faculty and to providing some funds for scholarships and fellowships for graduate level education (Crane, 1976; Jones, 1977). The vehicle for the in-service education was summer institute programs. Although these institutes began for college faculty, businesspeople in the high technology industries had firmly established that the weakest link in the supply of well trained scientists and engineers was the high school. Corporations such as General Electric, Westinghouse, and DuPont contributed substantial resources to analyze and then to develop high school science programs. The activities of these corporations, particularly General Electric, predated any similar involvement by NSF in the in-service training of high school teachers (Crane, 1976).

The combined efforts of the major science and technology businesses and corporations, along with the newly formed NSF, had the initial effect of stimulating the production of programs for the development of future scientists. This emphasis was very important in the years to come as NSF began to expand its efforts into the development of programs for the improvement of science education at all levels. The theme of "science for scientists," while laudable in the early programs to improve science education at universities and colleges, would soon prove inadequate in NSF's efforts at the precollege level.

In 1975 all funding for science education was withdrawn from the NSF budget by Congress, which had been NSF's only funding source and had always had reservations about NSF dictating a school district's curriculum. This control of curriculum was a major issue in the loss of financial support. In 1956, the chair of the House Subcommittee of the Committee on Appropriations, Congressman Thomas, raised a fundamental issue at the NSF appropriation hearings that to this day haunts NSF.

> Here is a pretty good place for me to say this. I have read your justification carefully. I was struck with two outstanding thoughts. One is that it is not going to be very long before you have a charge thrown in your face, and you are going to have a hard time defending it, that the National Science Foundation is trying to take over in this country the colleges and the universities and lay down the pattern and control through a federal bureau in Washington, the education and training of college students and scientists. (U.S. Congress. House. 1955)

The NSF sought from the beginning to guarantee that the principal drive, design, and growth of the precollege programs would come from outside the Foundation. By design, NSF's involvement in science projects was administrative only. Once a project was approved by a peer review panel, implementation of the program or grant project was the responsibility of the university, college, or institution that received the award. The assumption was that the individuals awarded the grants for basic research were experts in their respective fields and should not be held to strict guidelines for the direction of the goals of the research or be governed by strict timeframes for completion of the research.

The NSF applied the same administrative policy to its education grants as a way to avoid "the curriculum control" argument brewing in Congress. The policy was essentially to allow institutions or individuals awarded grants to make all decisions related to grant-supported activities. This explains why the impetus for starting the precollege science programs came from outside the

TABLE 2.1 Early Activities of NSF in Science Education Programs and Level of Support by Year

Fiscal year and type of program	*Number of* *Institutes*	*Stipends*	*Support by NSF*
1953			
College	2	42	$ 21,000
1954			
High School	1	26	10,000
College	3	71	40,500
1955			
High School	4	126	59,000
HS and College	2	(included in HS figures)	
College	5	173	75,000
1956			
High School	13	750	380,000
HS and College	5	(included in HS figures)	
College	7	450	235,000
1957			
High School	87	4,800	5,055,000
HS and College	4	(included in HS figures)	
College	5	300	278,000
1958			
High School	120	6,000	6,400,000
College	5	300	285,000
1959 and 1960			
High School	320	16,000	
College Institute	35	1,400	19,750,000
College Conference	20	800	2,200,000
Technical Institute	2	80	100,000
Elementary	10	450	500,000

Source: L. Crane, 1976, p. 74; see note 1 in this chapter.

Foundation. As already indicated, initially all the activities NSF sponsored in the area of science education were teacher institutes under the direction of practicing scientists, mathematicians, and engineers who, for part of their assignment, were also engaged in teaching. The intent of NSF was for the scientific community to teach teachers. In the beginning it was the teaching of college level science teachers, but in a couple of years the summer institute programs were expanded to include high school teachers (Crane, 1976; Welch, 1979).

Table 2.1 presents a summary of the Foundation's early activities in science education. It documents both the rapid growth in the allocation of funds for NSF teacher institutes during the first years of NSF and the initial focus on college level programs. Two factors are important to note from the data provided in Table 2.1. First, the support for high school teachers quickly outdistanced the support for college teachers. By 1957 the level of support was 19 times greater in favor of high school institutes; two years later there was another fourfold increase in the funds allocated for these programs—20 million dollars in the years 1959 and 1960 combined. The second factor of note is that by 1959 NSF was involved in the support of science education at all levels.

The most significant event at this time, which is not reflected in Table 2.1, was the Foundation's decision to fund the development of high school science curricula (National Science Foundation, 1975; Welch, 1979). It is significant because it would lead ultimately to NSF's involvement in the implementation of curricula in the schools. The efforts of supporting both curriculum development and curriculum implementation, which occurred over the next 15 or so years, made it difficult for NSF to show it was not mandating a national curriculum.

The reason for the involvement in writing curricula was simple: What good would the training of teachers accomplish if they were sent back to their classrooms to teach from outmoded curricula using outdated textbooks? Before we begin to dig more deeply into NSF's sponsorship of the science curriculum, which began in 1956, it is important to explain the reasons behind Congress' tenfold increase in funding from 1956 to 1957. Both the involvement in curriculum and the increased funding for science education programs occurred simultaneously, and it is hard to separate one from the other.

SPUTNIK AND THE RACE FOR TECHNOLOGICAL SUPREMACY

Accompanying the United States' interest in the development of programs to improve science research and science education was concern about how the Soviet Union was performing in these same arenas. During the years following World War II the Soviet Union had begun to engage in its "Iron Curtain" foreign policy. Very little information was getting into or out of the country without the direct involvement of government officials. However, around 1954 international competition from the Soviets in technology began to surface. Reports from the Soviet Union began to emerge that indicated that it was aggressively developing educational programs and industrial complexes to support science research and technology. Noteworthy among the reports was one by Nicholas DeWitt (1955) titled "Soviet Professional Manpower—Its Education, Training, and Supply."

National Science Foundation administrators used these reports to argue for increased funding in the operational budget, but for some reason the reports were not taken seriously by members of Congress. Members of Congress simply dismissed them as propaganda. This can be seen in the exchange in 1955 between Alan Waterman, director of NSF, and Congressman Thomas, chair of the Subcommittee of the House Committee on Appropriations.

> *Dr. Waterman:* At the same time there is every reason to believe that the U.S.S.R. has already surpassed us in the production of engineers and is on the way to doing so in the production of scientists.
>
> *Rep. Thomas:* I notice that you, 2 or 3 times, refer to that through your justification. How much is fact and how much is fiction? How did you arrive at that statement?
>
> *Dr. Waterman:* We have their 5-year plan, in fact, a succession of 5-year plans for education. We have the benefit of people who have been there traveling. There was a book written on the subject by an Australian who was attached to the Embassy there. I have a summary of information which came out in *Business Week* as the result of an interview with me. Recently the *Harvard Review* came out with a report which was abstracted, a digest of it, by Nicholas DeWitt in *Aviation Age,* which represents some of their findings. Of course, one cannot go into all sources of information by which this is received.
>
> We have to make a compilation of what we can get. Things like this are fairly clear, things such as are stated in this article by DeWitt. We have copies of this for members of your committee, if you like.
>
> *Rep. Thomas:* What is the basis of his information? People travel from this country into that country, but the bridle is put on them when they get there. They are shown the things they want them to see. I do not see how you can put too much credence in any of their propaganda.
>
> *Dr. Waterman:* I think we can count on this, Mr. Chairman, they have a larger population than ours, and they are evidently making a drive in education.

> *Rep. Thomas:* Do you have any information of what they are doing in the public school system? Do they have a public school system?
>
> *Dr. Waterman:* They do. We understand that 40 percent of the time of high school students is devoted to required science, that if they fail to pass two courses they are dropped. All along the way there is a selective process. . . . Every evidence is that they are doing the utmost to develop their science as a source of economic strength and a strength in defense. (U.S. Congress. House. 1955, p. 230)

The implied threat of Soviet superiority in 1955 did have an impact on the Appropriation Committee's funding decision for the 1957 NSF budget (see Table 2.1). Moreover, that implied threat, which Congress did not take very seriously, was viewed as a real threat on October 4, 1957. On that day the Soviet Union successfully launched and placed into orbit its Sputnik satellite and in so doing immediately earned respect from members of Congress as a technological competitor. Congress approved an additional emergency budget allocation of 9 million dollars in 1958 to be used for science education institutes, and in 1959 NSF earmarked a record 46 percent of its total budget for science education programs.

Science education had become a national priority. Major amounts of financial and human resources would be expended in the next 15 years. From this point forward, the United States would strive to establish the necessary machinery to ensure that it would reign supreme in the competitive world of science and technology. On one level the efforts paid off, and the United States has been successful in the preparation of advanced science researchers. On another level, however, it failed. How else do we explain the withdrawal of funds from NSF in 1975 and the antagonistic position taken by the National Academy of Sciences and Engineering in 1982? A partial answer to both questions can be found in the procedures used to design and implement the science curriculum.

DESIGNING CURRICULUM: SCIENCE FOR SCIENTISTS

Under the direction of practicing scientists, the first curriculum development grant was awarded to the Physical Science Study Committee (PSSC) in 1956. The position was taken that summer institutes would not have an impact on the teaching that occurred in schools if the teachers were using outdated textbooks and curricula. Over the next 10 years, the development of new curriculum materials grew rapidly. In 1964 the National Science Foundation was supporting seven elementary or junior high school science projects and five projects at the secondary level. Just two years later, the number had increased to 26, 19 in science and 7 in mathematics (see Table 2.2).

The PSSC project established the procedures all other curriculum projects would follow. The projects developed new science textbooks and laboratory manuals, and designed the equipment needed to carry out the curricula. Typically, project teams were composed of scientists, teachers, and administrators. It was clear from the very start that the scientists were in charge (Welch, 1979). Individual projects were directed by prestigious scientists, coordinated by advisory boards composed of prestigious scientists, and written by scientists.

The role of teachers and administrators was primarily to provide feedback to the scientist-writers. Once a curriculum draft was ready, it would be distributed to teachers in test classrooms across the nation. Based on the trial run of the materials, changes would be made. More often than not, however, teachers' feedback had very little effect on subsequent versions of a curriculum. The scientists were reluctant to accept changes to their science from schoolteachers (Jones, 1977, Welch, 1979).

A critical factor, then, in understanding the history of NSF curriculum projects in science is the dominant and decisive role members of the scientific community played in the development of precollege science curricula. This role had a significant effect in determining the focus of the curriculum—science for scientists.

Another important factor to consider is the speed with which the NSF-sponsored curriculum projects were developed and implemented. It was, in a word, phenomenal. By 1965, 50 percent of all physics students (200,000) were using PSSC materials, and 350,000 students were using the NSF-sponsored chemistry curriculum (CHEM Study) materials (Goodlad, 1966). By 1968, the estimated number of secondary school students using curriculum materials developed under NSF grants was 3,369,500 (Crane, 1976, p. 132). In less than a decade, science curricula were written and implemented. The nation was in a race, and a valiant effort was being made to ensure America's supremacy in science and technology.

In addition to establishing a model for the administration of subsequent curriculum projects, PSSC's general philosophy also became the philosophy of other NSF-sponsored curricula. The philos-

TABLE 2.2 Principal Curriculum Study Groups in Precollege Education

Subject	*Title*	*Sponsor*
Math	Committee on School Math	Univ. of Illinois
	Computer-based Math Education Project	Stanford Univ.
	Cambridge School Math	Educational Studies Inc. (ESI)
	Madison Project	Webster College
	Univ. of Illinois Arithmetic Project	ESI
Math & science	Minnesota Math and Science Teaching Project	Univ. of Minnesota
Physics	Physical Science Study Committee (PSSC)	ESI
	Harvard Project Physics	Harvard Univ.
	Engineering Concepts Curriculum Project	Comm. on Engineering Ed.
Chemistry	Chemical Education Materials Study (CHEM Study)	Univ. of California
	Chemical Bond Approach Project	Earlham
Biology	Biological Sciences Curriculum Study (BSCS)	Univ. of Colorado
General science	Secondary School Science Project	Princeton Univ.
	Earth Science Curriculum Project (ESCP)	Am. Geological Inst.
	Introductory Physical Science (IPS)	ESI
	Elementary School Science Project	Univ. of Illinois
	Science Curriculum Improvement Study (SCIS)	Univ. of California
	Elementary Science Study (ESS)	ESI
	Elementary School Science Project	Univ. of California
	Committee on Science Education	Am. Assoc. for the Advancement of Science
	School Science Curriculum Project	Univ. of Illinois

ophy stressed that teachers teach students how to operate as scientists, or, in the particular case of PSSC, as physicists. Learning by doing was the approach, and it soon became known among science educators as teaching science as inquiry. Thus, what we find is physicists writing for future physicists, chemists for future chemists, biologists for future biologists, and so on.

What is important to note here is that the chronological order of the curricula developed under NSF sponsorship—namely, physics, chemistry, and biology—was the reverse of science courses taught in high school. The significance of the sequence is that it established as the goal of science education students' attainment of the highly mathematical and logical final form version of science found in physics and chemistry. In Chapter 5 the parallels between this aspect of curriculum development and philosophy of science during the twentieth century will be explored. In each, guidelines established in physics became the standard for all the other sciences. Essentially, the purveyors of precollege science education were supporting a view of science that would come under serious criticism by philosophers of science. In another decade, the view would be rejected.

But even in 1958 there was concern about whether the type of science education being proposed by PSSC was what public education wanted or needed. Initial reactions by educational researchers to PSSC's curriculum were quite critical and called for the integration of philosophy, logic, statistics, and psychology. A particularly poignant issue was whether the teaching of the scientific method by teaching students—how to operate as scientists—was a significant educational objective (Easley, 1959).[1] This was an important question for two reasons. First, precollege science education programs would soon come to endorse the concept of teaching science as inquiry—learn science as scientists do science (Schwab & Brandwein, 1962). Second, the development and implementation of the curriculum materials—which taught students to operate and think as scientists—occurred at a time when basic concepts about how the growth of knowledge actually occurs were being drastically altered.

CURRICULUM IMPLEMENTATION: A POLICY SHIFT

Just as the decision to involve teachers in the summer institute programs led to the argument that these teachers would need updated curriculum materials, the decision by NSF to support the writing of science curricula led to the argument that the teachers would need training in how to use the new science curricula.

> The PSSC physics curriculum, as with many later NSF-sponsored curricula, was set up in such a way that the student learned by doing experiments, by making his or her own observations, and by generalizing on this first-hand experience. This curriculum required that the teacher be able to teach students how to operate as scientists in the classroom. In the case of the PSSC course the students were to operate as physicists. Since very few high school teachers know physics, much less know how to operate and think as a physicist must (even on an elementary level), the National Science Foundation found that it had to get involvd in teaching teachers how to use modern science curricula. (Crane, 1976, p. 65)

The involvement of NSF in curriculum implementation is critical. Not only did curriculum development and implementation grow to become a major component of the NSF educational effort in the years to come, but the decision to begin sponsoring programs for the implementation of curricula represented a major change in policy for the National Science Foundation. This policy change would, as previously indicated, eventually lead to charges that NSF was mandating curriculum which has historically been the dominion of local school boards.

In addition to the involvement in teacher institutes, NSF began to sponsor other efforts. One of the special projects in science education was the Cooperative College–School Science (CCSS) program. In 1959 this program was described as one for training elite scientists. Grants were administered to universities, which in turn selected secondary schools in their areas; together they planned and designed the best instructional programs possible for high ability students. Only the most able teachers were selected to participate in the program, and they were encouraged to enroll in summer programs of study at appropriate universities.

By 1963, however, an important format change had occurred in the CCSS program. It involved the new activity of postsecondary and secondary school personnel getting together to introduce the newly written NSF-sponsored curricula into one or more nearby school systems. The CCSS program had become the mechanism through which information about the new NSF science curricula could be brought to schools. In more blunt terms, it opened the door for the marketing of products. In the CCSS program before 1963, the training sessions for teachers were external to the curriculum being implemented. In 1963, however, training and curriculum implementation were aligned into a single program. Here is how one program was described.

> First, the teacher-participants will be given training, including laboratory work, during a summer phase so that they will have the necessary background material for working with one of the improved elementary science projects. The second phase consists of a coordinated academic-year program of weekly meetings, demonstration classes, and assistance with implementation of the elementary science materials. At the conclusion of the project, the school district will have adopted improved science curricula for the kindergarten through the eighth grade with trained teachers who understand how to use the materials. (Crane, 1976, p. 115)

However, by 1964 it was clear that curriculum implementation and not teacher training was the focus, and Congress raised questions about whether the shift of funds from institutes to the CCSS program meant that teachers were selected differently for the summer programs. Indeed they were. But more important, teachers were participating in sessions in which the science taught to them was the science they would teach to children. It was a watered down approach and often quite insulting to a person with a background in science. The instructors of the programs were also more often than not faculty from colleges of education, rather than scientists.

The shift in funds that occurred from 1962 to 1963 was large, a decrease of 6.7 million dollars to institutes and an increase of 5 million dollars to the CCSS program. The increase in emphasis at NSF on the CCSS program can be said to have had two effects.

1. There was a shift away from teachers and toward the schools and school systems, or away from people and toward institutions.
2. There was a shift in the control of precollege education programs away from scientists and toward educators.

The first shift was part of the curriculum and staff development needed to retool a school district from the old to the new. It was indeed necessary, but it took on an air of "follow the guidelines and all will be fine." The sessions with teachers became sessions of how to teach and not what they needed to know to teach. The second shift can be viewed as causally related to the first shift and is a clear reversal of National Science Foundation policy.

> From the beginning the Foundation has taken the position that since one of the principal weaknesses of science education was the fact that its content—the subject matter itself—was obsolete and dilute; teachers, books, and students were expending much valuable time, space, and effort upon inconsequentials and missing matters of importance. Hence, the Foundation attempted to secure the cooperation in a movement of national reform of the subject-matter experts, the practicing scientists, mathematicians and engineers who were also engaged in teaching. The idea has been that the scientific community should teach teachers, should help to give talented students scientific experiences and instruction beyond that normally available in their regular schooling, should actively work in writing and producing better instructional materials and programs and finally, begin to dig deeply into fundamental research on the teaching-learning process. (Crane, 1976, pp. 121–122)

The year 1962 was the beginning of the curriculum implementation efforts by the CCSS program, the same year the National Academy of Sciences and Engineering cited as the beginning of the downfall of science education. From 1962 to 1972 the CCSS program experienced an eightfold increase in funding. During that same period, funding for precollege teacher institutes was cut in half. Frank Press' assertion that science education began to slip in the year 1962 takes on new meaning when we recognize these shifts. Thus, the inference that seems to have been made by members of the National Academy is that the failure of the first NSF effort to sustain improved precollege science instruction was not the failure of scientists, but rather was caused by the implementation strategies used by educators, schools of education, and school systems.

Is this a fair assessment? It will be argued in Chapter 3 that the purveyors of the early precollege science curricula—the scientific community—effectively ignored relevant developments in the history and philosophy of science. Consideration of such developments could have assisted teachers in making curriculum decisions and in the long term in becoming better teachers of science. In the rush to design and implement new science curricula, the directors of the early NSF curriculum projects did not do their homework (Connelley, 1972). Although the science content was accurate and the support materials designed for use in the classrooms were excellent, the curriculum research had not been done. It has been shown that the intentions of the curriculum developers were lost in the transmission to teachers and students (Connelley, 1972). Analyses of the curricula indicate they were misdirected to address a style of teaching science rather than a style of learning appropriate to science (Novak, 1977).

There is one other set of factors that explain the need to reexamine the health of science education in our schools today. In the 30 years between 1952 and 1982 our society became very dependent on science and technology. Whereas the concern in 1952 was educating future scientists to produce technology, starting in 1982 the concern became educating all citizens to participate in the highly scientific and technological world produced by the previous generation. There has been a shift in priorities in precollege education. Now science is deemed important for the worker on the line and the mid-level manager, as well as for the decision makers. The contemporary problems in science education are not so much with the training of research scientists. Rather the problems today are educating all students to be scientifically literate in the twenty-first century. The contemporary solutions must also attempt to prepare a work force that depends more and more on technological advances.

It is a challenge that will ultimately be addressed by individual teachers working in individual classrooms. One of the most important lessons learned from our first comprehensive involvement in precollege science education is that the classroom teacher is the key (DeRose, Lockard, & Paldy, 1979). Although the infusion of new curriculum and support materials makes a difference, it materializes only when teachers know how to use and, perhaps more important, maintain the curriculum. As the students change, as the knowledge base of science changes, as the processes of science change, and as materials need to be replaced, so too must the design of instructional tasks be changed.

Therefore, teachers will need to be competent and effective decision makers in the selection, sequencing, and implementation of instructional tasks. They will need guidelines to assist them in making these decisions. Teaching is a profession characterized by an environment in which the important variables—students, teacher, subject matter—are always changing. To repeat a theme from Chapter 1, precollege science teachers will continually engage in activities concerning the retooling and reexamination of the science curriculum.

NOTES

[1] In 1959 the *Harvard Educational Review* dedicated a major portion of one issue (Vol. 29, No. 1) to reporting the papers delivered at a symposium on the Physical Science Study Committee. Participants included noted educational researchers and science educators Jack Easley, Leo Klopfer, and Fletcher Watson, among others. Given the changes in science education occurring at Harvard at that time, the reactions were in some regards predictable. Harvard University was participating in the redesign of basic science education for its undergraduates. James Conant, president of the University, contributed significantly to the development of a classic work in science education, *Harvard Case Histories in Experimental Science* (1957). This work sought to integrate concepts from the history of science into the science curriculum. The spirit of science education at Harvard at the time is perhaps best demonstrated by the fact that it was while engaged in the preparation of materials for a case study that Thomas Kuhn (1970/1962) began to develop the ideas for his seminal work *The Structure of Scientific Revolutions*.

Chapter 3

Rethinking Our View of Science Education

The science curriculum modifications in the 1950s sought to accomplish two things: (1) upgrade teachers' and students' knowledge of what was known in the world at that time, and (2) instill in students a keen interest in science. The principal goal of the NSF curriculum projects was to produce a new generation of scientists. Although this goal is highly commendable, it will be shown that the Foundation's efforts to improve the quality and quantity of science instruction came at a time when the very character of science was being redefined. Simply stated, the change was an abandonment of the view of science that held that inquiry proceeds exclusively from statements of observation to statements of theory, and that scientific knowledge is added generation after generation, thereby accumulating more accurate information over time.

In the 1950s philosophers of science, supported by the work of individuals in the new academic discipline of history of science, began rejecting the idea that observations and theories could be treated separately. The building block view of scientific knowledge was also being dismantled. In their place developed a view that argued that science is better conceived as an activity in which (1) theoretical commitment determines observational standards, and (2) replacement, substitution, and even outright abandonment of knowledge claims, in lieu of additive growth, are more accurate descriptors of the growth of scientific knowledge.

The purpose of this chapter is to place contemporary science education into a recent historical perspective. Worldwide, the activities in science education in the 1950s and 1960s were the most ambitious attempt ever made to alter the character of instruction in a K–12 discipline (Hodson, 1985). Examining the historical context in which science education curricula and practice emerged from these two pivotal decades enables us to identify the positive aspects of those efforts and, more important, to identify the things that were left out.

This chapter describes the central role of theories in understanding, learning, and teaching science. Adopting a view of science that does not focus on the development and justification of scientific theories would contribute to the use of one type of instructional strategies for teaching science in the classroom. A science teacher who knows and understands the role theories play in science will employ a very different set of criteria for selecting and designing instructional tasks, sequencing them, and determining the most important content in science lessons.

If there is one tool that a proper view of science provides a teacher, it is the skill to evaluate scientific knowledge claims, and, thereby, to see that such claims are not all equally important. Some theories are more important than others, some scientific facts are more significant than others, and some experiments are more crucial than others. Along these same lines it is wrong to assume that scientific theories and scientific facts have equal status. It is also incorrect to judge and reject older theories of science (such as the geocentric view of the universe, solid earth theory, and caloric theory of heat) as less eloquent or rigorous, and therefore less scientific, than contemporary theories. The establishment of these claims is one of the many contributions historians of science have made to our understanding of the discipline of science. That both past and present contributions of science should be recognized is also critical for teachers of science to understand.

THE RISE OF HISTORY AND THE ROLE OF THEORY IN SCIENCE

Chapter 2 noted that the charter for the development of the National Science Foundation was a manifestation of World War II. Also pointed out in that chapter was the pivotal role this august institution played in the design of precollege science education programs. To further understand the activities of the National Science Foundation and the products of the various curriculum project committees, it is necessary to broaden our analytical perspective. Specifically, we need to consider developments in two closely related and important disciplines that use the concepts and processes of science as their objects of investigation. The two disciplines are philosophy of science and history of science.

Philosophers, in general, seek to understand the uniquely human process of thinking and the factors that influence thought processes, such as ethics and logic. Philosophers of science focus their attention on matters addressing the growth of knowledge. Inasmuch as science is an activity that seeks solutions to problems and explanations of questions, philosophers of science strive to explain (1) the processes that describe the activities of inquiry and determine the form of knowledge (called epistemology), and (2) the status of the products of these processes (ontology). Examples of epistemological and ontological questions are presented below.

Epistemology
What counts as an observation?
What counts as an accurate test of a hypothesis?
How are scientific theories generated?
What is the relationship between observations and theories?

Ontology
Are scientific principles and theories true or merely useful tools to guide inquiry?
Are the facts, principles, and concepts of science real or just human descriptions of the world at this point in time?
Are our scientific descriptions and explanations of nature truly accurate or do we impose them on nature?

Our interest in philosophy of science begins with the start of the twentieth century. A number of new important and sweeping theoretical statements about the world emerged in the nineteenth century. Major new theories took root in biology, physics, and chemistry. Charles Darwin, Andrew Wallace, and Gregor Mendel changed biology with their theories of evolution, natural selection, and genetics. Ludwig Boltzman, William Thomas (Lord Kelvin), and James C. Maxwell were some of the players in the dynamic changes in the field of physics. The kinetic theory of heat, the laws of thermodynamics, and principles of electromagnetism contributed to very new views of atoms and wrought fundamental changes in concepts of energy and matter. The chemical atomic theory emerged from the collective activities of John Dalton, Joseph Priestly, Antoine Lavoisier, Avogadro, Stanislao Cannizzaro, and Dmitri Mendeleev.

What confidence could thinkers at the beginning of the twentieth century have in these new theoretical speculations? A select group of scientists and philosophers of science in and around Vienna, Austria, said, "Very little." They began to take issue with theoretical statements built from or based on unobservable data. More exactly, the criticism and discussion focused on whether science could establish a solid foundation on such ephemeral theoretical speculation. In particular, these critics took issue with the use of atomic models on the grounds that (1) they wrongfully made real things that could not be observed, and (2) they could not guide research because of how complicated and inconsistent they had become (Brush, 1988). Foremost among the physicists criticizing the atomic theory was Ernst Mach, who wrote:

> Thereby we suppose that things which can never be seen or touched and only exist in our imaginations and understanding can have the properties and relations only of things which can be touched. We impose on the creations of thought the limitations of the visible and tangible. (Brush, 1976, p. 286)

The concerns were both epistemological and ontological (refer to the lists above). The reaction was to promote a philosophy of science—positivism—that stressed the importance of science proceeding from observable evidence to accurate predictions. Thus, great care was taken to develop logically consistent rules outlining how theoretical statements could be derived from observational statements. The intent was to create a single set of rules to guide the practice of theory justification. For the philosophers of the so-called Vienna Circle, the objective was to develop one singular form for judging all theoretical statements in science. Crucial elements of this form were empirical observations and logic. Hence, the labels for this brand of philosophy include logical positivism and logical empiricism.

One important element of logical empiricism is the separation of observations from theories. This separation is known as the observational/theoretical distinction. We still find the influence of this view of science in science textbooks that encourage students to observe or "discover" natural phenomena and scientific concepts without any understanding of the fundamental concepts or principles needed for seeing and discovering. The second important element of logical empiricism is the role of logic. Logic is applied only to the testing or justification of theories; it is not believed to have any hand in the creation of theories. History of science helped to dismantle both of these components of logical empiricism, and by 1969 scientific theories had become the cornerstone of philosophy of science (Suppe, 1977).

Thus, for logical empiricists a theory of science was considered a strong theory only if its theoretical statements could be logically justified by observational statements. The members of the Vienna Circle, whose motivation was to establish confidence in the knowledge claims being generated by scientists, attempted to set down the dynamics of "the scientific method" as defined for physics and applied to all other scientific disciplines.

From the very beginning, however, the Vienna Circle's attempt to bolster the need for empirical observations, to make theory dependent on observation, and to establish a singular logical form for science inquiry met with resistance. Scientists did not cooperate; the speculative theoretical thought of the 1800s did not just continue, it expanded. Theoretical speculation dictated observation. Some very speculative theories emerged in science during the first several decades of the twentieth century; for example, Einstein's theory of relativity in physics and the collective theories of quantum mechanics, Alfred Wegener's theory of continental drift in geology, and the great synthesis of Mendelian genetics with Darwinian natural selection in biology.

Theoretical speculation was alive and well and contributing to the growth of science. Attempts to hold such speculation down were feeble, and constant revisions were made to the philosophy of empiricism; over time, those revisions accounted for the increased role of theoretical statements in the growth of scientific knowledge (Grandy, 1973). From the beginning of the twentieth century to the post-World War II years, continual reconsideration of the role of observation and theory in science resulted in several brands of empiricist philosophy. In chronological order the major empiricist philosophies were positivism, logical positivism, and hypothetico-deductivism (Losee, 1980).

It is the last of these that science teachers would immediately recognize as the standard scientific method. It involves

1. Selecting a hypothesis
2. Conducting observations
3. Collecting data
4. Testing the hypothesis
5. Rejecting or accepting the hypothesis

The hypothetico-deductive method of justifying scientific knowledge claims (that is, testing hypotheses) had one very important characteristic in common with the early twentieth-century philosophies—observations were still considered to be mutually exclusive of theories, and logic could be applied only to the testing of theories, not to the discovery of theories. The application of logic only to the testing of theories persists in our science programs. What is being argued here is that with the aid of history of science, it is possible to construct a logic for the discovery of scientific ideas. We will discuss how this logic of discovery can, in turn, be used to format instruction.

The relationships between observation and theory, and between the testing and discovery of knowledge claims became two of the most important issues in philosophy of science in the 1950s and 1960s. They are represented by the following questions:

Is it possible to have observations free of theory?
Do commitments to theory affect the measurement and interpretation of data?
How do scientists develop theories?
Is there a single preferred form for theories?
Is the development or alteration of theoretical ideas rationally constructed?

As philosophers struggled with answers to these and other questions about the growth and form of scientific knowledge, a new academic discipline was emerging that came to have a significant impact on philosophy of science. Each academic discipline can point to a particular point in time when the activities of practitioners firmly established both the methodological procedures and research agendas that others would follow. For example, calculus was devised by Newton in the sixteenth century, modern chemistry nomenclature was born in eighteenth-century France and England, structural geology and principles of stratigraphy were developed in nineteenth-century England, and mineralogy had its start in nineteenth-century Germany. The new discipline we are concerned with is history of science, which was developed in both Europe and the United States in the twentieth century.

The credit for establishing history of science as a scholarly pursuit in the United States goes to George Sarton. Sarton, a Belgian by birth and a scholar of ancient and medieval science, started the first journal solely devoted to historical studies of the development of science—*ISIS*. He also developed the first American Ph.D. program in history of science at Harvard University, which awarded its first degree in 1947 to I. Bernard Cohen.

Thomas Kuhn, author of the most acclaimed book in history of science, *The Structure of Scientific Revolutions* (1970/1962), remembers those early days.

> I first became interested in the history of science during 1947, at which point I was a graduate student two years from a Ph.D. in theoretical physics. Four years after that introduction, still having taken no course in the history of science, I offered my first to a small group of Harvard undergraduates. And five years later still I moved to Berkeley, where I had been invited to set up a program in my new field. No trajectory of that sort is imaginable today, for history of science has in the interim become a profession. . . . Other signs of that transformation are not hard to recall. When I was working my way into the field, there were only half a dozen people employed to teach it in college and universities in the United States, and no two of them were at the same institution. (Kuhn, 1984, p. 29)

Besides laying the foundation for making history of science a scholarly activity, Sarton also firmly established a style of doing history of science. For Sarton good history of science meant going well beyond merely cataloguing the chronological order of the successes of science. Sarton put in place a set of methodological guidelines that sought to understand the choices scientists made in the pursuit of scientific explanations. Again, Kuhn writes, "I was drawn . . . to history of science by a totally unanticipated fascination with the reconstruction of old scientific ideas and of the processes by which they were transformed to more recent ones" (1984, p. 31). Thus, the paths not taken were as important to the historian as the path ultimately selected. Following World War II, when the world was settling back into its normal paces, many scholars turned to the study of history of science. The impact their works had on philosophy of science was as revolutionary as the revolutions of science they were reporting. (A list of influential works in the history of science is provided at the end of this chapter.)

These critical histories of science discovered that the growth of scientific knowledge did not grow without disruption, upheaval, or alteration of central ideas. Close scrutiny of historical events in science indicated that science was better characterized as a discipline in which dynamic change and alteration were the rule rather than the exception. The view of science as an inductively logical process—a process of moving from empirical fact to the development of scientific theory—was not supported by these historical studies either. The new methods for writing history of science and the new findings from such analyses provided historians and philosophers of science with evidence that the rigid form of science advocated by logical empiricists and logical positivists did not exist. Rather, these histories of science revealed that all aspects of science—its standards, meanings of terms, application of methods, and theoretical forms—progress through stages of development (Thackray, 1980). What emerged was the view that theoretical commitments determined much of what was done in science.

At times theory development leads to novel ideas, and a wholesale redirection of scientific commitment and activity takes place. Thomas Kuhn called such changes "scientific revolutions." Western European civilization has seen only two complete scientific revolutions. The first occurred in the period 1500–1800 as a result of the work of Nicolaus Copernicus, Galileo, Descartes, Newton, Lavoisier, and others. The second scientific revolution took place in the nineteenth century and is defined as the "successful quantification of the Baconian sciences" (Kuhn, 1977, pp. 218–220). Brush (1988, pp. 4–5) expands Kuhn's definition to include the revolutionary developments in physics (1895–1925), biology (after 1859), psychology (1895–1920), and nuclear physics (1945). Some of these developments are identified in Chapter 4.

With its national sudies; discipline studies; science and religion studies; science, medicine, and technology studies; philosophy, psychology, and sociology of science studies; and great people studies, history of science has had a significant impact on philosophy of science. These historical studies provided very new perspectives for philosophers of science. It became increasingly apparent, more than ever with the aid of history of science, that what was observed, measured, evaluated, or hypothesized in science was done with strong theoretical commitments. In other words, not only did an observational/theoretical distinction not exist in science; if anything, it seemed that theory determined observation. Norwood Hanson (1958) put it this way: What we see is determined by what we know. This same general opinion has carried over into our views of how children learn science—what is being learned is affected by the existing knowledge base of the learner.

Unfortunately for science education, conclusions and views such as these were unavailable to science educators and scientists preparing new curriculum materials for precollege science programs in the 1950s. Similarly, the new views emerging from philosophy of science were not fully considered by individuals working on NSF science curriculum projects. The result is that at the same time that science education was adopting new perspectives for teaching science as an "inquiry into inquiry," the basic definitions of inquiry were being altered (Duschl, 1985). The next section examines this critical period of mutually exclusive development. We are, in many respects, still employing science instructional strategies that do not reflect the richer understanding we have of science today, an understanding made possible by inquiries conducted by historians, philosophers, and sociologists of science.

CHANGING THE FOCUS OF SCIENCE EDUCATION

It is significant for science teachers to note that the contributions from history and philosophy of science were being developed at the same time that the National Science Foundation was involved in funding the development of K–12 science curriculum projects. The first of the numerous curriculum development projects funded with NSF money was the 1956 Physical Science Study Committee (PSSC) project conducted at M.I.T. The PSSC project established the procedures all other curriculum projects would follow (Welch, 1979). Details of these were outlined in Chapter 2, but it is important to recall here that the general approach was for teachers to teach students how to operate as scientists. In educational parlance, it was called teaching science as inquiry. In the case of PSSC, it meant students should learn to inquire and operate like physicists. However, performing science and understanding science are, as discussed in several chapters, very different activities requiring very different sets of cognitive tasks.

During the same period of time (1956–1966) in which various science contents were being revised to produce curricula that would instruct students how to operate and think like scientists, the prevailing ideas about what it meant to think like a scientist were being challenged and changed by historians and philosophers of science. The concept of making science instruction an inquiry into inquiry (Schwab, 1962) was and still is a good idea. The inquiry approach in science firmly established the role of the laboratory and the doing of science by children. The problem is that in the 1950s and 1960s inquiry was limited to learning how to test knowledge claims. Today scientific inquiry means so much more.

Following is the list of processes of science (Mayor & Livermore, 1969) adopted for the Science–A Process Approach (S–APA) curriculum directed by the American Association for the Advancement of Science:

Basic Processes	*Integrated Processes*
Observing	Defining Operationally
Measuring	Stating Hypotheses
Using Numbers	Reading/Making Graphs
Space and Time Relationships	Controlling Variables
Communicating	Designing Experiments
Predicting/Inferring	

These 11 processes are representative of the final form brand of science presented in most curricula. The basic processes make up a large portion of the K–6 science curriculum, while the integrated processes are intended to be used in the 7–12 curriculum. The process approach is a carryover of the view of science defined by empiricism that begins with observations and then proceeds logically to the formulation of scientific theories. It is a view of science provided by practitioners (physicists) working with neat problems that can be precisely measured and with hypotheses that can be tested with true controlled experiments. Missing from this view, however, are two important processes of science—explanation and evaluation.

The significance of explanation and evaluation is that each represents a product of scientific inquiry determined by the theoretical commitments an individual or a community of individuals—in this case scientists—adopts in the construction of a personal world view. That is, what counts as an explanation for you may not be satisfactory to me, and vice versa. What makes an explanation unacceptable or what makes it appear to be correct is the criteria we employ in our evaluation. The criteria science employs to set standards—the standards of measurement, of selecting important research questions, of designing an experiment, of accepting the outcomes of an investigation—play a very important role in the growth of science.

The value to science teachers of understanding how science explains and evaluates is that psychologists have discovered that these processes are not dissimilar from how individual learners develop a personal scientific view of the world. Succinctly stated, the growth of knowledge in the discipline of science is emblematic of the growth of knowledge in an individual learner. The dynamics of this position will be explored in more detail in Chapter 6. For now, however, let us begin to appreciate that knowledge growth, or cognitive development, is best characterized as a process in which concepts are continually replaced.

This idea of constructing knowledge becomes a central element of our discussion. Understanding, at least to some extent, that scientific knowledge is constructed and reconstructed is crucial to a full appreciation of the arguments herein. Such construction, though, is not always built on a sound foundation nor is it always designed to employ the best building materials. Science is an activity that, when examined carefully, reveals many false starts, the use of faulty logic and wrong assumptions, and the employment of investigative methods rooted to theoretical commitments. It is an activity in which knowledge claims are, at best, always tentative. An important challenge we face as science educators is how to present science as a rational activity if the product of that activity is tentative knowledge. The solution lies in teaching both faces of science.

TWO FACES OF SCIENCE REVISITED

Today with our knowledge of the history of science in hand, our understanding of inquiry in science includes not only processes for testing knowledge, like those identified in S–APA, but also processes for generating knowledge. There is a logic that can be applied to scientific discovery. There are, then, two faces or characterizations about the nature of science.

1. Science as a process of justifying knowledge—what we know.
2. Science as a process of discovering knowledge—how we know.

The first characterization dominates contemporary science education. Knowledge claims, facts, hypotheses, and theories are typically taught for the contribution each makes to establishing modern knowledge. How such knowledge came to exist is, as the logical empiricists proposed, treated as a non-issue. The consequence is that an incomplete picture of science is presented to students. Learners are provided with instructional tasks designed to teach what is known by science. A large part of this instruction involves teaching students processes that justify what we know.

This is the first face of science, the profile established by the testing of knowledge claims and represented by our contemporary view of the world. This type of instruction is characterized as instruction that seeks to improve students' *scientific knowledge*. This is the testing face of science.

What is presently missing in our science curriculum are instructional units that teach about the other face of science—the how. It is important for students to know how we have come to believe one view about the world (a 4.5 billion-year-old earth) instead of another view (a 10,000-year-old earth), or why we employ one method of investigation for data collection (satellite infra-red imagery) over other methods (aerial photographs). It is not enough to simply tell students the facts. To learn science meaningfully requires much, much more. What is missing in science curricula is the other face of science—the discovery face.

Teaching about the "what" without teaching about the "how" runs the risk of making science instruction incomplete. Kilborn (1980) has used the term *epistemological flatness* to describe science curriculum materials or instruction strategies that do not give a complete picture of the concepts being taught. Too often, he argues, science instruction is taken out of context and presented without the critical background material necessary for an understanding of the meanings or transitions of science. One important context to consider is the historical context. Others include the political and social contexts in which science functions. The intermingling of science with technological advancements and social conditions cannot be denied.

Using instructional materials that do not provide a context for exploring how we have come to know, or why this or that is considered an important question to investigate, is to run the risk of teaching a science class that does little to convince students that science is an activity in which change is a normal and rational part of the growth of knowledge. When instruction focuses exclusively on what is known, teachers are providing students only part of the story. Students are left to fill in the gaps themselves. Yet recent research on how students construct personal views of the workings of the world clearly suggests that students make frequent mistakes when filling in the gaps (Novak & Gowin, 1984; Osborne & Freyberg, 1985; Wittrock, 1986). More important, these gaps also seem to prevent students from succeeding in science at higher grade levels. Essentially, science becomes inaccessible to many young learners.

An alternative is to recognize the second face of science—the discovery of science face. Here the intent is to design instructional units that seek to develop students' *knowledge about science.* Knowledge about science, as opposed to scientific knowledge, is knowledge of why science believes what it does *and* of how science has come to think that way. The distinction between scientific knowledge as a curriculum objective and knowledge about science as a curriculum objective is based primarily on the exclusion and inclusion, respectively, of history of science and of the important role of theory development in science. The key to understanding science is to understand the important and diverse roles theories play in science.

In Chapter 4 our discussion turns to scientific theories. More specifically, several characterizations of scientific theories will be presented to establish a foundation we can use to design instruction that employs both faces of science. Bringing science education programs and science teaching into the mainstream of thinking about science requires perspectives provided by history and philosophy of science. Each of these two disciplines makes science an object of investigation. The respective contributions of scholars working in these two disciplines deserve serious consideration by science teachers and science curriculum writers. Rationally reconstructing the events and lines of reasoning that have contributed to our modern view of the world is an exciting alternative for science education to take. At a minimum it ought to be included in one instructional unit on the development of the major theories of the disciplines we teach.

Theories are misrepresented in our science classes; often they are treated as simple statements of definition to which relatively little time is allocated. Evidence of this can be found in the debate on teaching evolution. Public opinion in certain states has persuaded textbook publishers to insert disclaimers of the following type: Evolution is only a theory. Scientific theories are complex and deserving of the same comprehensive effort toward learning about them as went into constructing them.

Scientific theories represent our best reasoned beliefs about the world and thus become the standards against which other knowledge is tested. Giving scientific theories the status and attention they deserve in our science curriculum represents a dynamic and complex challenge for science teachers. Chapter 4 begins to outline how the structure of scientific theories can provide pragmatic guidance for teachers in selecting and sequencing instructional units. Theories, as we will see, have a rich and descriptive language all their own. The procedures adopted for testing and generating theories represent guidelines we can employ to facilitate student learning in science.

APPENDIX

The following books are recommended as an introduction to topics in the emerging field of history of science:

Albritton, C. C. (1963). *The fabric of geology.* Reading, MA: Addison-Wesley.

Bronowski, J. (1974). *The ascent of man.* Boston: Little, Brown.

Brush, S. (1988). *The history of modern science: A guide to the second scientific revolution, 1880–1950.* Ames: Iowa State University Press.

Cohen, I. B. (1985). *Birth of a new physics.* New York: Norton.

Durbin, P. (Ed.). (1980). *A guide to the culture of science, technology, and medicine.* New York: Free Press.

Geikie, A. (1905/1962). *The founders of geology.* Reprinted. New York: Dover.

Gillispie, C. C. (1960). *The edge of objectivity.* Princeton, NJ: Princeton University Press.

Hall, A. R., & Hall, M. B. (1964/1988). *A brief history of science.* Reprinted. Ames: Iowa State University Press.

Holton, G. (1978). *The scientific imagination: Case studies.* New York: Cambridge University Press.

Kuhn, T. (1977). *The essential tension: Selected studies in scientific tradition and change.* Chicago: University of Chicago Press.

———. (1970/1962). *The structure of scientific revolutions,* 2nd ed. Chicago: University of Chicago Press.

Laudan, R. (1987). *From mineralogy to geology: The foundations of a science, 1650–1830.* Chicago: University of Chicago Press.

Losee, J. (1980). *A historical introduction to the philosophy of science,* 2nd ed. New York: Oxford University Press.

McKenzie, A. E. E. (1973/1988). *The major achievements of science.* Reprinted. Ames: Iowa State University Press.

Mayr, E. (1982). *The growth of biological thought.* Cambridge, MA: Harvard University Press.

Reingold, N. (Ed.). (1979). *The sciences in the American context: New perspectives.* Washington, DC: Smithsonian Institution Press.

Schneer, C. (1960/1984). *The evolution of physical science: Major ideas from earliest times to the present.* Lanham, MD: University Press of America.

———. (Ed.). (1969). *Toward a history of geology.* Cambridge, MA: M.I.T. Press.

Thackray, A. (1980). History of science. In P. T. Durbin (Ed.), *A guide to the culture of science, technology, and medicine.* New York: Free Press.

Toulmin, S., & Goodfield, J. (1962/1982). *The architecture of matter.* Chicago: University of Chicago Press.

———. (1965). *The discovery of time.* New York: Harper & Row.

Chapter 4

The Status of Theory in Science Education

Close critical examination of activities associated with the growth of scientific knowledge indicates scientific theories are the cornerstones of science. Scientific theories represent our best reasoned beliefs about the world around us. They are, in a word, explanations—a synthesis of facts, aims, and methods of science.

It is also clear that theories progress with time through a type of rite of passage. New theories are always greeted with skepticism, and rightly so. Of these only a few are recognized as valid and pass the rigorous tests of the scientific community. Still fewer prove to be central core theories of science—explanatory schemes that direct the work of future generations of scientists. It is a characteristic of this progression that new theories replace older theories. How does this replacement take place? What criteria are used to guide the replacement of one explanation of science by another?

The status of theories in science is a type of paradox or enigma of scientific knowledge. On the one hand, we must recognize theories as the standard bearers of science. As such, select theories play a major role in determining research questions, methodologies, and standards for evaluating (rejecting or accepting) investigative results. On the other hand, history of science clearly establishes that the scientific community oscillates between times when there is a consensus about the status of a theory and times when there is a dissensus. There are times when scientists agree and times when scientists agree to disagree.

At any given time, each scientific discipline can be described by its core theories. What is relevant for science teachers to realize is that each of the science subjects taught in grades 7–12 has since the 1800s experienced major changes in the theoretical dynamics of the core discipline that makes up the curriculum (see Table 4.1). There exists, then, a number of case studies that demonstrate the

TABLE 4.1 Examples of Theory Change in Science

Discipline	*Century*	*Theoretical Core*
Geology	18th	Neptunism—rocks formed by water
Geology	19th	Plutonism—rocks formed by fire and water
Chemistry	18th	Phlogiston explains burning; molecules (H_2O) are pure substances that have properties (spirit) of animate objects
Chemistry	19th	Oxygen explains burning; elements (H,O) are pure substances that have properties of periodicity
Physics	17th	Planetary motion is circular and at constant velocity; weight and mass are combined
Physics	18th	Planetary motion is elliptical with changing velocity; weight and mass are separate
Biology	18th	Life classified/grouped according to habitats (i.e., things that swim, fly, and so on); diversity of life explained by spiritual laws
Biology	19th/20th	Classification by morphology; diversity of life explained by mechanistic laws

consensus (agreement) and dissensus (disagreement) that occur in science among practitioners.

It is a major task of science education to present instructional activities that explore the shifts from point of view A to point of view B. This presentation is a two-way street, however. It is a path that can be described by both consensus and dissensus among scientists. That is, decision paths exist that describe both the construction of new theories and the dismantling of extant theories. Consequently, a critical element in understanding science is to understand the steps taken in scientific theory decision making. The benefit for teachers is that by placing the decision-making path into a historical framework, it becomes possible to reconstruct, at least partially, the consensus and dissensus activities that contributed to our present day understanding. I, for one, am convinced that many students are perplexed not so much by what science believes but rather by how science has come to see and describe the world in that particular way. Examining the path and the decision making of communities of scientists allows us to begin to appreciate the rational evolution of scientific knowledge.

In this and the following chapter we will detail characteristics of scientific theories that can be used as procedural tools by science teachers to format instruction. More exactly, we will examine

1. Mechanisms for theory change
2. Criteria for classifying theory types
3. Procedures for testing and evaluating theories
4. Competing characterizations of the role of scientific theories in science
5. The important relationships between observation and theory.

IMPORTANT CONTENT AND LEGITIMATE DOUBT

This book began by paying homage to the formidable foe every science teacher faces—time. Teachers of science assigned to teach junior high and senior high school science classes must confront the challenge of how to cover all the objectives set forth in school district curriculum guides. It is a significant challenge, which should not be taken lightly. For example, in one study (Duschl & Wright, 1989) it was discovered that the science programs in a senior high school had more behavioral objectives listed in their curriculum guides than there were numbers of days in the school year. The teachers in these schools saw to it that students covered—not learned or understood—the material.

The position taken in this book is that it is better to explore a select set of science concepts, methods, aims, and processes in depth than to take the smorgasbord approach so common in science books and science curricula today. Taking a depth rather than a breadth approach is critically important and has been endorsed by science education, mathematics education, and educational psychology researchers (Linn, 1987). The basic argument is that learning occurs when students understand the relationship between or among concepts. In science as well as in other disciplines, such relationships become quite complex in a short time. Fortunately for science educators, we are provided with final form versions of these relationships among concepts. The final forms are our scientific theories.

The challenge we face in teaching final form versions of scientific knowledge to our students is it involves both a building process and an undoing process. In Chapter 1 we pointed out that research on students' understanding of basic science concepts has found that many students harbor very wrong ideas about how the world works. Students seem to start science instruction with a faulty foundation of concepts. Learning in science is described as a process in which old ideas, concepts, and meanings are replaced by new ones. The task for science teachers, then, is to design instruction that assists students in changing their novice views of the world into the more sophisticated scientists' view of the world.

Changing from one point of view to another: Does that sound familiar? Presenting science as an activity that supports revision and substitution of scientific knowledge claims helps create a learning environment that supports students' attempts to revise their ideas. When teachers make theory change and revision an object of their planning and instructional design, they contribute to the development of a classroom climate and a science curriculum that accurately portray and promote knowledge growth in science. Examining the salient characteristics of theory growth and theory change provides teachers with new criteria to help make decisions concerning the selection and implementation of instructional tasks. Good teaching is always preceded by good planning. Good planning anticipates students' needs and requires a thorough grasp of the subject.

Deciding that science education should change presents us with the problem of determining what in the curriculum goes and what stays. For us the decision will be guided by scientific theories, just as in the real world of science. We will ask ourselves, as we are often asked by our students, to provide answers to two very important questions:

1. What is the most important content?
2. When is it appropriate to question science or the claims of scientists?

The solutions reside in coming to terms with the context that will be selected for teaching science. When the context highlights the evolution of concepts, methods, and aims that contribute to the growth of scientific theories, the task is greatly simplified.

FIGURE 4.1 The Goal-of-Science Hierarchy

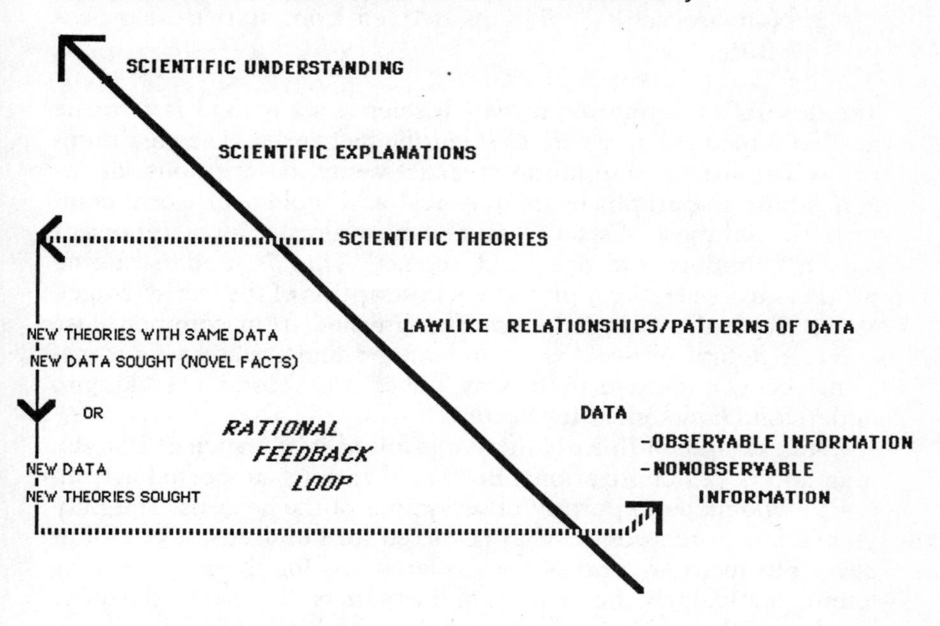

THE GOAL OF SCIENCE: SOUND SCIENTIFIC THEORIES

The goal of scientific inquiry is the development of scientific understanding. A critical step toward achieving this goal is the development of sound scientific theories. Regardless of the scientific discipline, theories are the result of scientific activities that seek to provide explanations of phenomena. To better appreciate the role theories have in the enterprise of science, it is helpful to represent and outline the role theories have in a goal-of-science hierarchy.

Figure 4.1 is a schematic representation of the relationships among scientific data, lawlike, empirical relationships or patterns of data, scientific theories, scientific explanations, and scientific understanding. The bottom of the hierarchy is where scientific inquiry begins; the top represents the goal of such inquiry. Understanding how scientific inquiry proceeds through the hierarchy is a study of the nature of science.

All science investigations begin with the collection of data. It is at this level that science first establishes a concrete connection with the natural world. The data collected may be of two forms:

1. Observable—capable of being seen or measured.
2. Nonobservable—relations derived from statistical probabilities.

The level of investigation at which science seeks data is also the level at which the majority of scientific facts exist. The vast number of catalogues containing measurements, descriptions, drawings, and photographs of the physical and biological world comprise the database of science. But such catalogues alone represent only a commonsense notion of science. That is, understanding what is involved merely provides a description of the world (Nagel, 1960). Scientific knowledge is distinguished from commonsense knowledge first by seeking to understand and explain *why* something occurs and exists in the way it does, and second by seeking to understand *how* something occurs.

An example of this distinction is found in the ancient Babylonian and Greek civilizations. Both civilizations supported astronomers who made important observations of the heavens. The Babylonians kept records of what occurred for hundreds of years but never put forth any models or explanations for the motions they found, particularly the motions of the planets. Greece, on the other hand, has the dubious claim of being the birthplace of science. Why? Because the Greeks were the first to speculate on causes and explanations for observed events. In short, they proposed theories. Facts in themselves do not provide an understanding of the world. It is the relationships among the facts—the theories, models, and explanations we put forth—that constitute our representations of understanding.

Toward the goal of providing scientific explanations, then, two levels of representation must be successfully achieved. The first is the development of lawful or lawlike relationships among the facts; the second is the development of generalizations that can be drawn from the first set of relationships. It is at this second representational level that scientific theories are created. An example will help to clarify the point.

Earthquakes have always been part of human history. That the earth shakes occasionally is an undisputable fact. Knowing that earthquakes occur is commonsense knowledge; knowing why they occur requires explanations based on scientific knowledge. Since the middle of the nineteenth century, no fewer than five explanations for the cause of earthquakes have been put forth (Duschl, 1987). Among the explanations are the gravitational force of the moon, changes in barometric pressure, isostatic rebound of mountains, moving plates of crustal rock, and rising gas from the mantle.

Not until the middle of the twentieth century were enough data on epicenters and foci of earthquakes gathered to allow scientists to discover several previously unknown patterns about the location, depth, and distribution of earthquakes. The first pattern or lawlike relation is that earthquakes occur on the ocean floors. The second pattern is that the depth of ocean-floor earthquakes changes with location. In the middle of the oceans earthquakes occur only close to the ocean-floor surface, and foci never exceed depths of 70 km. (The focus is the point below the surface where the earthquake actually occurs. The epicenter is the surface location of the earthquake directly above the focus. One kilometer is equal to .6 mile.) Near the margins of the oceans, however, earthquakes are found to occur in bands of increasing depth: shallow, ≤ 70 km; middle, > 70 km and ≤ 300 km; deep, > 300 km. Curiously, the maximum depth of earthquakes seems to be about 700 km. The third pattern is that the distribution of earthquake epicenters on the earth's surface, both on the ocean floor and on continents, is not random but follows linear paths of earthquake belts.

Prior to World War II there was little knowledge of the surface profile of the ocean floor. The development of SONAR, however, created a new observational tool for seeing the ocean floor, which led to the discovery of data patterns two and three (above). The fourth pattern is that the ocean floor was found to have long, high mountain chains throughout and deep trenches near the margins. The discoveries about the distribution of earthquakes and the topography of the ocean floor laid the groundwork for yet one more discovery. It was found that the depth at which earthquakes occur below the surface aligns with specific features of the ocean floor. For example, the shallow earthquakes in the middle of the ocean correlate exactly with the mid-ocean mountain ridges. The shallow, middle, and deep bands of earthquakes are found only near the deep ocean trenches. This match of earthquakes to landforms ultimately contributed to the theory of sea-floor spreading, which, in turn, led to the development of the theory of plate tectonics (Takeuchi, Uyeda, & Kanamori, 1970; Wilson, 1976).

The epicenter data and the SONAR data represent separate sets of data at the first level of the hierarchy. Next each set of data was used separately to construct relationships that outlined the distribution patterns of earthquakes and the profile of the ocean floor.

This is the second level of the hierarchy. But only through the synthesis of both sets of data were explanations of why earthquakes occur and why they occur where they do provided. The synthesis represents the third level of the hierarchy. The explanation, of course, is the theory of plate tectonics.

One lesson here is that facts derive meaning from theory, not vice versa. It was mentioned previously that over the last 100 years scientists have proposed several explanatory models for the causes of earthquakes. In the late 1800s earthquakes were explained by the gravitational effects of the moon, analogous to the moon's effect on the tides. In the early 1900s changes in barometric pressure were identified as a partial cause of earthquakes. At the same time the isostatic rebound of mountains—floating adjustments of rock—explained earthquakes. A more recent rival to the plate tectonic theory is the proposal involving rising methane gas from the mantle. In each explanation the fact that earthquakes occur remains the same; descriptively there is no change. What changes is our descriptions of why and how they occur.

As our example demonstrates, scientific knowledge is susceptible to frequent and, at times, dynamic changes. Here we saw how changes in technology (SONAR, seismograph) prompted changes in the data at the base of the hierarchy. As science educators we face the challenge of convincing students that change is a normal element of the growth of scientific knowledge. At times the changes are minor, at times major. There is no reason to believe that science will not continue to alter the patterns and theories that make up our explanations of the world. Educationally, the issue is the chance that students will wrongly interpret science as an irrational activity or line of inquiry. Knowing, in general, how science goes about the process of self-correction addresses this problem.

Changes in scientific understanding typically occur in one of two ways. First, different interpretations from a single set of data may be generated at either the lawlike relations level or the scientific theory level. An example is polar wandering vs. drifting continents to explain changes in the direction of rock magnetism. Another is genetic influence vs. environmental influence to explain behavioral characteristics of humans. The existence of competing ideas among scientific explanations is common and will be discussed in more detail later in the book.

The second, and the more common, mechanism for altering scientific understanding, is to effect a change of the data. Such changes are more frequent because of (1) the sheer numbers of scientists, (2) technological advances that allow for new observational forms of data, and (3) changes in the aims of science as an outgrowth of societal problems. With each technological advance (the microscope, telescope, x-ray, computer, to name but a few), new sets of data emerge that create new scientific horizons. With each socially important problem (space exploration, AIDS research, weapon advancements for wars), new teams of researchers turn their attention to seeking solutions. Again, new data emerges.

These two mechanisms of change—new theories from existing data and new theories from new data—are represented in Figure 4.1 as the rational feedback mechanisms of science. In trying to understand the rationality of scientific change, one discovers that when radical changes occur in scientific explanations, there is often a chain of reasoning connecting the two different sets of criteria. The path of change from the original View A to a newer View B frequently, but not always, follows a rational evolution (Shapere, 1984). It is this rational evolution that has importance for science teachers. These models can be used to determine the most important content.

Needed details about the issue of knowledge growth must be added to our teacher decision-making arsenal. The frequent changes in science that are characteristic of the growth of knowledge have led many people to refer to theories as merely ideas or sets of speculations no more important than the facts. The example of developing explanations for earthquakes demonstrates that this view of theories is wrong.

The frequent successes of science also contribute to the false assumption that scientific knowledge is a cumulative, ever building, and accretionary process. Of course, this assumption is to be expected. The notion of theory as an idea is a common application of the term in our language. Furthermore, it is hard to deny that science has improved not only our understanding of the world but also the world we live in. Consider the list of twentieth-century scientific discoveries presented in Figure 4.2 and how each has shaped our lives (*Science 84*, 1984).

Scientific explanations in the form of scientific theories do not miraculously appear. On the contrary, scientific theories are rigorously and arduously drafted. It is the pressure of time in the classroom that smooths over the presentation of the path taken in the growth of scientific knowledge. When we neglect to present science as a process of revision and substitution of knowledge claims, we run two risks. One is developing in students the perception that

FIGURE 4.2 Major Twentieth-Century Discoveries in Science

1900–1919

Leo Baekeland discovers resin used to make nylons and plastics.
Alfred Binet develops IQ test that has defined intelligence for 70 years.
Albert Einstein formulates relativity, helps create quantum theory, and establishes the reality of the atom.
Karl Landsteiner discovers blood groups, enabling successful blood transfusions to be done.
Karl Pearson introduces statistics to the world of scientific decision making.
Lee De Forest invents vacuum tube and opens airwaves to radio and TV.
George Shull develops hybrid corn from crossbreeding experiments, making it possible to feed the world.

1920–1939

Medicine is transformed from an art into a powerfully effective applied science with the discovery of antibiotics.
Raymond Dart shows that Africa brought forth the first human species.
Nuclear fission is discovered in radioactive materials.
Edwin Hubble deduces the big bang theory from data of an expanding universe.
DDT inspires both agriculture and the environmental movement.
Vladimir Zworykin invents the television camera, solving the problem of converting light to electricity and electricity to light.

1940–1959

Russell Marker synthesizes, from Mexican yams, hormone used in birth control pills.
Computers are developed to break Nazi codes and soon open up new worlds of data representation and analysis.
Chlorpromazine and lithium empty mental hospitals and change our understanding of mental illness.
Bell Laboratories invents the transistor and introduces solid-state electronics.
Watson and Crick unravel the secret of DNA and alter our views of life.
Laser light discovery opens up new windows.

scientific-knowledge growth is governed by the addition of new ideas, facts, and theories to old ones. The second risk is portraying science as an activity in which scientists always seem to agree or have a consensus. We as science teachers must ensure that these misapplications of theory and misinterpretations of science do not take root in our science classes.

Nothing could be further from the truth than to characterize science as an activity that only adds new knowledge to existing knowledge. It is certainly hard to dispute that science has contributed to a better understanding of how the world works. Yet historians of science have shown us that all too frequently theories, methods, and goals of science are replaced and or simply abandoned. To appreciate the rigor with which scientific explanations are developed, it is necessary to probe well beyond the mere facts and success lists of science. The theory of plate tectonics, for example, took 60 years to bring to full development. It was rejected in its initial form as continental drift, put to sleep for 30 years due to objections by believers in a solid-core earth, revived by physicists employing remanent rock magnetism data, and eventually adopted as sea-floor spreading of oceanic plates.

When examined broadly, science appears to be an enterprise in which bold inventive developments and amazing new discoveries are the rule. When examined carefully, however, science is an activity in which agreement or consensus is balanced with disagreement or dissensus. New ideas do not magically emerge. New ideas typically replace old established ideas. When one idea is being constructed, another, competing idea is usually being dismantled. There are false starts, dead ends, and temporary success with methods, with aims and goals, and with models, theories, and explanations. Recognizing that scientific change is complex and involves both the construction and the dismantling of explanations suggests two very important attributes of the structure of scientific theories.

1. All things in science do not have equal status. Science places preferred status on theories and theoretical models. The unequal status applies not only to theories over facts, but also to levels of theories.
2. All things in science change with time. The line of reasoning involved in such change involves complex sets of consensus and dissensus decision-making activities. To claim a knowledge of science is to claim an understanding of the mechanisms of theory change.

These two attributes are vital for developing teacher decision-making guidelines determined by subject matter. Thus, each is presented at length: The first, levels of theories, in the remainder of this chapter; the second, mechanisms of theory change, in Chapter 5.

LEVELS OF THEORIES

Theories, like everything else in science, have a developmental history. All theoretical explanations that attempt to generalize across data sets and explain lawlike empirical relationships have a beginning. It is absurd to think that only useful ideas have ever been proposed in science. The past is no different from the present in this respect. Certain explanations come to have scientific merit (for example, Copernicus' heliocentric model of the solar system), while others begin as hoaxes and remain hoaxes (such as Von Daniken's grandiose claims that intelligent life visited the earth centuries ago and left the pyramids and other ancient architecture as evidence of the visit). In either case, though, each is introduced as a new idea that is met with criticism and skepticism. The task that faces students of science is how to distinguish the crank ideas from the truly scientific.

One way to distinguish among scientific theories is to use criteria associated with problem solving. L. Laudan (1977) employs problem-solving ability as a set of criteria to differentiate the status of theoretical-knowledge claims. For him there are two types of problems—empirical problems and conceptual problems—and two types of theory-generating activities—progressive research activities and degenerative research activities.

Empirical problems are best thought of as the data sets, facts, and lawlike patterns a theory needs to explain. The more patterns a theory accounts for, the more progressive it is. Conceptual problems are the controversial issues or debates a theoretical model either encounters or proposes. A theory degenerates if it encounters either empirical or conceptual problems; it progresses if it avoids such problems. Thus, in general, the more empirical problems (past, present, or future) a theory solves and the more conceptual problems it avoids, the more progressive the theory and the research tradition that spawned the theory. If, on the other hand, a theory is unable to solve empirical problems (past, present, or future) or is fraught with conceptual problems, then the theory and the research tradition are degenerative. Two examples will help clarify these distinctions.

Predicting the positions of planets was an important problem in early astronomy. The backward loop, or retrograde, motion of Mars, however, defied explanation and was an empirical problem that needed to be solved. Any theoretical model of the heavens that could provide a solution to this problem would be a more progressive research tradition.

Early astronomers were able to predict quite accurately the position of the planets Mercury, Venus, Mars, Jupiter, and Saturn; the sun; and the moon. What is intriguing about this is that their theoretical model for the organization of the heavens placed the earth at the center. The observations, measurements, and predictions they made were quite good and worked within this geocentric model. But the retrograde motion of Mars was a problem. How was it possible to explain the brief backward looping motion of the planets?

Ptolemy solved the problem with the theory of major and minor epicycles—a wheels within wheels model of planetary motion in which the motion of planets rode on small circles (epicycles) attached to the larger circular orbit (deferent). It is analogous to riding a bicycle. As one peddles, the circle made with the foot travels front to back and then back to front, and so on. The motion of the bike is always forward, but the motion of the foot is simplified to a back-and-forth motion. Extending the analogy to planetary motion, when the planet is moving in the smaller circle in the same direction as the motion of the larger circle, normal west-to-east direct motion is observed. But when the motion of the smaller circle is opposite to the motion of the larger circle, east-to-west retrograde motion is observed. With the earth at the center of the universe, this model adequately explains planetary motion, although the predictions of the planets' positions that it provides are imperfect.

Stimulated by the need for more accurate predictions of planetary motion, Nicolaus Copernicus was quite willing to alter the geocentric model. What, he surmised, would happen to the predictions of planetary positions if the planets and the earth were placed in orbit around the sun and if the earth rotated on an axis? Recalculating the data employing these new motions yielded the same lawlike patterns of planetary motion as the geocentric model. But Copernicus found that he could calculate the siderial period (time of one revolution around the sun) for each of the planets and that the distance of the planet from the sun was directly proportional to its siderial period.

With this distance/speed relationship established, Copernicus was able to provide an alternative explanation for the retrograde motion of Mars. Rather than retrograde motion being an actual physical motion of the planets involving epicycles, Copernicus was able to argue from calculations of the positions and orbit velocities of the earth and Mars that the retrograde motion of Mars was a type of optical illusion. As the earth moved more quickly through its orbit than Mars, Mars would appear against different backgrounds of stars as seen by an observer on earth. Copernicus had solved an empirical problem.

But there was dissensus among astronomers. Copernicus' heliocentric model was not ushered in as a viable alternative to the geocentric model because it failed to explain another problem. If, it was argued, the earth both rotates and revolves, then an observer on earth should see a parallax shift of stars when the planet is at opposite ends of the orbit. No shift is seen, so it does not follow that the earth is in motion. (Parallax shift can be simulated by a person placing a finger six inches in front of his or her nose, first looking at the finger with the right eye and observing the position of the finger against the background, and next, without moving the finger, closing the right eye and opening the left. The finger will appear to move against the background. Blinking back and forth accents the effect.) Copernicus could not overcome this conceptual problem. Not until Galileo argued for the effect of the enormous distance to the stars was the conceptual problem of parallax shift resolved.

Yet another conceptual problem with a moving earth concerns its rotation. Copernicus found that by rotating the earth, the motion of the sun could be explained. Furthermore, by combining earth rotation with earth revolution, he was able to explain the change in elevation of the sun's path through the sky. The knowledgeable reader will recognize this as the explanation of the seasons. Copernicus had solved yet another empirical problem.

Copernicus' theory of a rotating earth met with the same type of conceptual problems as did the Greek scholar Aristrachus' first proposal of a rotating earth. Why do we not feel a wind in the face? Why when an object falls does it fall directly below and not to the left or right of the release point? If the earth rotates, then we should observe these phenomena. Aristarchus' theory withered on the vine because his proposal was purely conceptual and did not account for any empirical problems.

Copernicus' theory, on the other hand, met with greater support by virtue of the cadre of solutions it provided to outstanding empirical problems—seasons, planet positions, retrograde motion, and precession of the equinoxes. Having established a successful empirical problem-solving record for his theory, Copernicus had only to wait for the removal of the conceptual problems that faced the heliocentric theory of the solar system. Moving the earth out of its position in the center of the universe held significant consequences for the religious views of the time. It was only through the collective work of scientists like Galileo, Kepler, and Newton—individuals who were persuaded by the empirical power of Copernicus' theory—that the conceptual problems were eventually resolved and dismissed. Unfortunately, this waiting period far exceeded Copernicus' lifetime.

We find in this synopsis of theory change examples of how empirical problem-solving ability and conceptual problem avoidance influence theory acceptance. We see also how the status of a theory changes with time and with the ability to solve problems. The ability of one theory (*Ta*) to explain/solve empirical problems of a rival second theory (*Tb*) is, at a minimum, necessary for the first to gain favor over the second. But if, as the goal-of-science hierarchy suggests, *Ta* suggests the existence of new, yet-to-be-discovered data and this data is observed, this provides a very compelling argument for ranking *Ta* over *Tb*.

Imre Lakatos (1970) referred to the prediction and discovery of new observational data from a theory as "novel facts." If, he argued, the new theory *Ta* could explain all existing empirical problems and also predict the existence of facts not suggested by the rival theory *Tb*, then that was a compelling reason to accept *Ta* as a better theory than *Tb*. Teachers who are able to identify the novel facts of their discipline during instruction are again adding a very useful element to their arsenal of curriculum selection and sequence decision-making skills.

Several examples in science support Lakatos' idea of the existence of novel facts in the growth of knowledge. Consider Einstein's prediction of the curving of light by gravity in his theory of general relativity. Observations of starlight traveling past the sun during a solar eclipse indicated he was right. Next consider Mendeleev's prediction of the existence of yet-to-be-found elements for his periodic table. From his rows and columns of elements, he was able to predict the approximate atomic weight and atomic number of elements that would fill the blanks in his chart.

From biology, consider the prediction about horses that was suggested by the theory of evolution. Othniel Marsh found a series of fossil horses in the geologic record that seemed to display a gradual change in foot bone structure. The fossil record had preserved the existence of four-toed horses, three-toed horses, and on to single-toed horses. When Thomas Huxley was in America in 1876, he had the chance to view Marsh's collection and recognized that a direct line of descent had been found. Several weeks later while delivering a paper, he predicted that a five-toed horse would be found. Two weeks after the prediction Marsh found such a horse.

The final example of theory anticipating observational data is from geology. J. Tuzo Wilson predicted that, if the theory of sea-floor spreading was correct, a new type of rock fault had to exist in which movement on opposite sides of the fault would "transform" from opposite direction movement to same direction movement. The offset character of the mid-ocean ridge bending and winding along the ocean floor demanded or required such a fault if the idea of moving plates was correct. Subsequent observations of the mid-ocean ridges proved him correct in his prediction of this novel fact.

The identification of empirical problem-solving power, conceptual problem avoidance, and novel facts assists us in theory ranking. It also assists us in employing another useful and simple scheme for presenting the developmental nature of scientific theories to students. Dutch (1982) proposes that scientific theories can be thought of as classifiable in either center, frontier, or fringe regions of science. This idea of regions of science is very similar to and fits nicely with another contribution of Imre Lakatos (1970). Lakatos argued that once theories (*Ta, Tb, Tc . . . Tn*) in a discipline had been assigned a general pecking order, they could then be assigned to either the hard core or soft core of a discipline. The metaphor he employs is that of a ball with an outer soft and inner hard substance (see Figure 4.3).

The outer soft core serves two functions. First, it protects the inner core from wild or untested theories. Second, it is the testing or proving ground for new theories. To combine L. Laudan's and Lakatos' ideas, only theories with high empirical problem-solving power and low conceptual problem-solving disputes will move closer to or into the inner core of a discipline's theory foundation.

The inner core is the hard core where the basic foundation of the theory is found. Here is where the strongest ideas that are most highly regarded by scientists exist. Whereas movement of theories can be into and out of the outer core, theories move only into the inner core. If the theories of the inner core are found to be inadequate and subsequently abandoned, then the ball must be discarded. An example is the central position of the earth in the geocentric theory of the universe. Once you give up on the earth being at the center of all motion, then it is a small step to reorganize all

FIGURE 4.3 Ball metaphor for classifying and ranking scientific theories

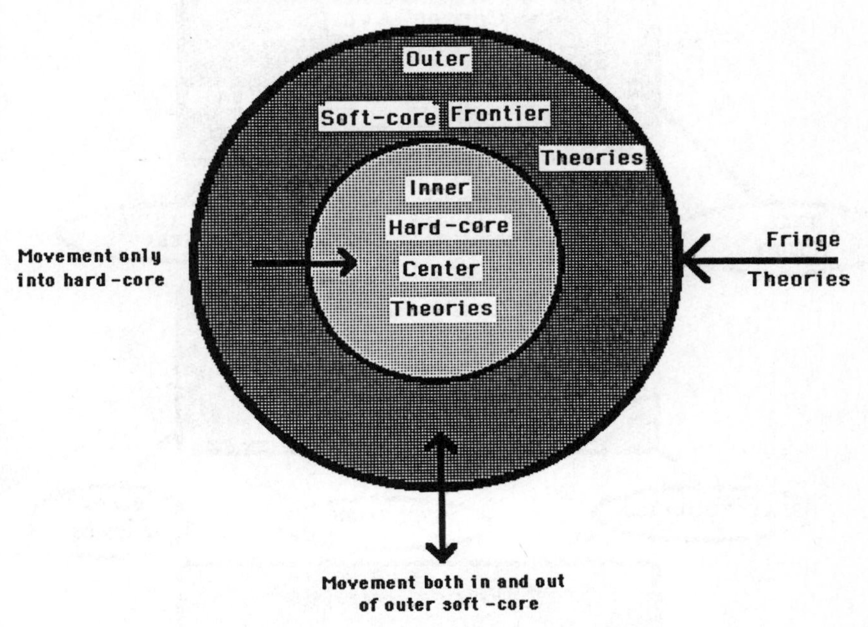

other elements of a model for the solar system (for example, Copernicus' solution of the retrograde problem).

Another way of representing the developmental nature of scientific theories to students is shown in Figure 4.4. Here we have extended the metaphor of the ball to apply Dutch's more educationally functional scheme, and have added a leather cover to the ball. The inner core represents the central theories, the outer core the frontier theories, and the cover the fringe theories of science. Theories at the center level, indicated at the top of the figure, are those explanations that are securely part of the mainstream of science. They have no rivals; that is, the scientific community does not entertain alternative explanations. Furthermore, to suggest that a central idea is faulty and should be replaced is to also suggest, as in Lakatos' hard core, that the entire set of concepts in a scientific discipline may be based on erroneous information. Examples of central theories are the theory of relativity, the kinetic theory of heat, the laws of thermodynamics, Kepler's laws of planetary motion, the cell theory, Newton's laws of motion, and the theory of periodicity of elements.

FIGURE 4.4 Levels of Theories

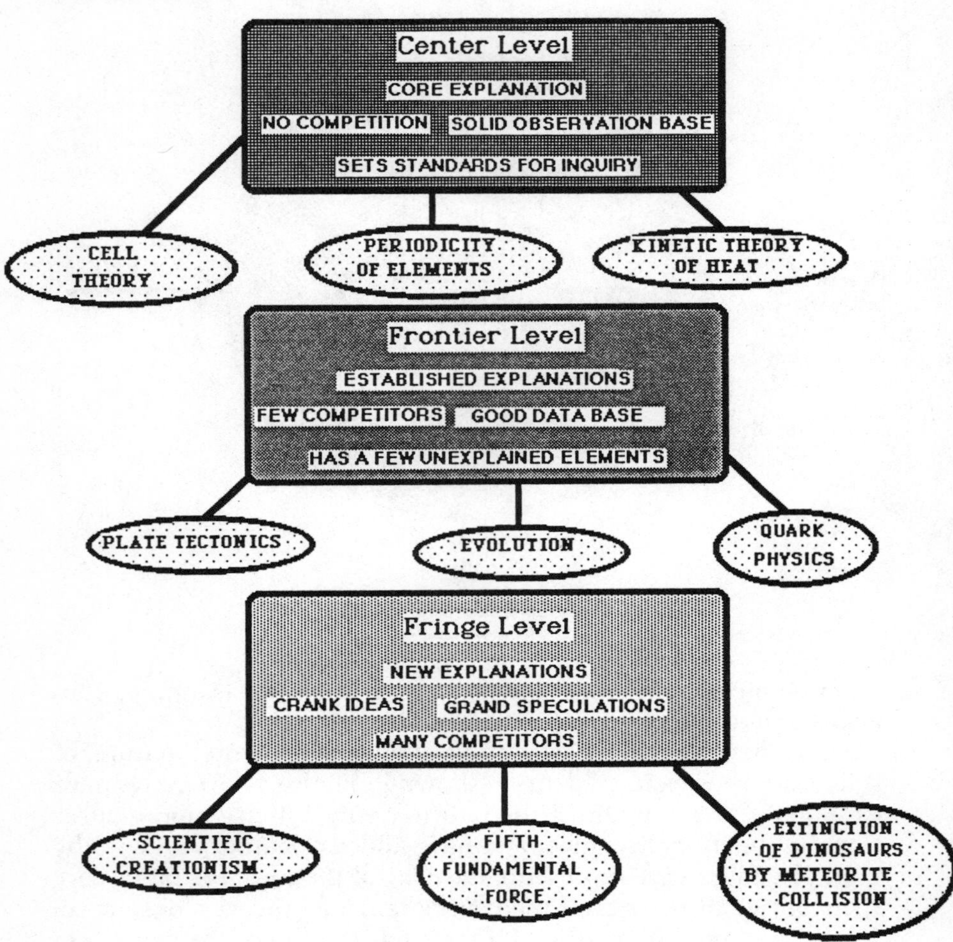

The second level is the frontier level of scientific theories. These theories are also part of the mainstream thinking of a scientific discipline. They are highly confirmed, based on sound scientific evidence, and endorsed by scientific communities for their ability to explain patterns of data and as a basis for predictions. Within these theories, however, there exist yet-to-be-resolved inconsistencies. Or, these theories have yet to fend off rival explanations; frontier theories do not stand alone in a discipline like central theories. Thus, frontier theories can be incomplete, but this does not seem to prevent scientists from employing the explanations to guide research. Clearly we see this with children's use of explanations. Putting theories into practice before total justification has been established is one enigma of science and has been labeled by philosophers the "underdetermination of theories" problem.

Examples of frontier theories are the theory of sociobiology, the big bang theory, quark physics, the theory of evolution, and the theory of plate tectonics. Each of these theories either is incomplete or has a rival theory that can be considered an equally viable alternative explanation. Evolution and plate tectonics suffer from a lack of certainty about mechanisms. Sociobiology and evolution seek to explain some of the same patterns of behavior in life. Quark physics is still searching for theoretically postulated particles, which may be experimentally identified when larger, more powerful cyclotrons, such as the Texatron, are built.

Taken together, central and frontier theories make up the bulk of any discipline's explanatory power. These theories set the standards—the hard core—for scientists. In turn, the theoretical standards determine or guide all other activities in science. Theories establish the meritorious research questions, the acceptable methodologies for experiments, the criteria for counting something as observable, the criteria for either rejecting or accepting the results of an experiment as evidence, and, ultimately, when new theoretical standards are necessary. Teachers of science should be well versed in the central and frontier theories that make up their discipline. In addition, they should be able to explain why these theories have come to hold such lofty positions. The criteria outlined above apply nicely.

The beginnings of theoretical explanations are not very glamorous, though they may be wild. The entry level of scientific theories is at the fringe of science, the cover of the ball. It is here, and here alone, that the off-the-wall ideas and potentially powerful scientific theories come face to face. Here is where crank ideas and sound theories can be subjected to analysis and the crank ideas can be rejected without disrupting an established body of scientific knowledge. Fringe theories are highly speculative and often lack confirmation. Some explanations will be revolutionary and eventually will attain the two higher levels. Others, for any number of reasons, will become the crank ideas and hoaxes of science.

Von Daniken's (1970) theory about intelligent life visiting the earth thousands of years ago is a fringe theory. So too are theories of parapsychology (sixth senses), scientific creationism, biorhythms, and Velikovsky's idea that a large comet was expelled by Jupiter about 1500 B.C. These ideas are considered to be crank theories because each paints an incomplete picture of the pattern of information it hopes to explain. In other words, there are compelling reasons to doubt the legitimacy of the knowledge claims being put forth. Foremost among the reasons are a lack of solutions to empirical problems and a plethora of conceptual problems.

Another new theory presently at the fringe but with the capability to be revolutionary concerns the extinction of dinosaurs. Scientists have proposed a twin star theory—our sun and another star—to explain the periodic extinctions that occur in the geologic record. The gravitational force between our sun and its companion star pulls our solar system and hence our planet in a galactic orbit. This orbit, the theory maintains, takes us to a region of the galaxy where there are large concentrations of meteors. It is hypothesized that an impact of one of these meteors with the earth—the site is held to be west of Greenland—was so large and so violent that it put large volumes of dust into the atmosphere, altered the climate of the planet, and led to the global destruction of the dinosaurs. This idea has been extended to explain the magnetic reversals of the earth.

> New theories, some highly speculative, have linked past impacts to a wide variety of key events in the Earth's history. The theories suggest that impacts caused reversals in the Earth's magnetic field, the onset of the ice ages, the splitting apart of continents 80 million years ago and great volcanic eruptions. . . . The impacts may also have played a major role in the evolution of life. Although scientists still debate the theory that climate changes wrought by massive collisions wiped out the dinosaurs, some evolution experts now suggest that such impacts may have caused numerous wrenching turns in the history of living species. (Sullivan, 1988, p. C1)

If the theory is true, then one should find some type of deposit of dust. It has been found (Alvarez et al., 1980). If it is true, then there should be some regular pattern to the extinctions in the geologic record correlating to the planet passing through the meteor zone. The pattern has been found. In spite of these promising results, however, the theory is a fringe theory with many gaps. It is at face value an extremely wild idea, particularly to science students of any age. It becomes one more theory among many about the possible scenarios for the elimination of the dinosaurs. But it has promise. Helping students see why the Alvarez theory has promise while the von Daniken theory does not is an important element in science education.

It is in the fringe region of scientific theories that confusion, if any, and debate about the directions of science are found. It is also at the fringe that excitement brews. My experience in the classroom indicates that students are informed about and excited by fringe theories—the wilder the better. What I am proposing is that these fringe theories, old and new, can serve as a type of tool to format instruction.

With new explanations, it is possible to outline criteria used to justify (solve empirical problems/avoid conceptual problems) and to generate (relevant research questions/established data collection devices/proposed by whom) the knowledge claim. It is easy to evaluate the status of the knowledge claims made by a fringe theory because the theory is typically pushing on conventional ideas in science. Being able to sift through the theory to extract the evidence, and the methods used to reach the theoretical conclusions, and to identify the aims of the proposal opens up a broad range of possibilities for science instruction. It seems that a doable goal of science education is to develop students' ability to evaluate the status of scientific knowledge claims. A feature of scientific literacy, not typically included in a definition of scientific literacy, is the ability to assess the degree of legitimate doubt attached to a scientific theory.

Being able to recognize valid revolutionary scientific theories from crank scientific theories of today and of the past is not as simple as it may seem. All theories of science—the central and the frontier theories—began on the fringe of science as new explanations or new models. Examining the progression of a few theoretical explanations from the fringe to the frontier and ultimately to the center is, I maintain, an important exercise for students to complete. All new explanations are treated initially as crank explanations; some, however, emerge from the pack to take their place as revolutionary explanations.

History of science enables us to ascertain what it is about the development of a theory that distinguishes it from other explanations of science. In short, the final form of scientific theories can be tested, and the developing form of scientific theories can be mapped. Taken together, scientific theories can be evaluated on two levels: (1) a context-of-testing level, and (2) a context-of-discovery level. Logic and empirical data supports the first level. History and sociology support the second level. Taken together, they provide a rational basis for change in scientific theories. Indeed, a fundamental axiom of science, as mentioned earlier, is that all things in science are not equal.

In Chapter 5 we explore in more detail some models that seek to describe and explain theory change in science. We have touched on that topic in this chapter through Lakatos' and Laudan's ideas of theory evaluation. It is necessary, however, to provide a more detailed analysis, if we are to apply it to curriculum selection, sequence decision making, and classroom practice planning.

Chapter 5

The Restructuring of Scientific Theories

In Chapter 4, basic characteristics of scientific theories were introduced. Among them were levels of theories, novel facts, and progressive and degenerative research programs. Our purpose in this chapter is to extend the discussion of theories to provide a richer framework for judging the status of scientific theories and for describing the processes of theory change. The extended discussion is needed because as science teachers we face a dilemma when we begin to make scientific theories an object of instruction or instructional decision making. When we attempt to establish criteria for judging theories and for describing theory change, we encounter the risk of portraying science as an irrational enterprise.

In Chapter 1, a distinction was drawn between the processes of science which address the origin and evolution of scientific-knowledge claims, and the processes of science which are associated with the gatherings of evidence and establishing the validity and reliability of scientific-knowledge claims. Designing a science curriculum to stress only one of these processes of science will necessarily present only a partial picture of science. We have learned that it is false to assume that the knowledge claims of science have equal importance or equal status among scientists. A challenge we face, then, as science teachers is how to determine what knowledge is most important.

Presently, teaching and learning models in science education—such as the learning cycle and the generative learning model—emphasize psychological bases for learning science. Psychological principles are important, but other factors need to be addressed in a comprehensive curriculum. Eisner (1985) suggests that there are five basic orientations a curriculum can have.

1. Curriculum as technology
2. Personal relevance

3. Development of cognitive processes
4. Social adaptation and social reconstruction
5. Academic rationalism.

Science education over the past 30 years has emphasized, to varying degrees, orientations 1, 3, and 5. Orientation 1 focuses on strict systematized planning of meaningful goals assessed in terms of behavioral objectives. It is safe to say that this orientation is the dominant orientation found in the majority of our schools. In this orientation, the decision what to teach is not made by the classroom teacher. Orientation 3 was the dominant focus of the 1960s elementary science curriculum projects sponsored by the National Science Foundation: Science–A Process Approach, SCIS, and ESS. Here, through carefully designed instructional strategies, the emphasis is on intellectual development and the cultivation of cognitive processes. In this orientation it is important to create a learning environment in which the student is an active agent in the process of learning.

Orientation 5 can also be found in the secondary science curriculum projects sponsored by NSF. Here the focus is on informing students about the special patterns of thought that apply to a discipline. In the case of science, students learn what it means to think like a biologist, chemist, geologist, or physicist. This orientation is preferred by liberal education proponents. Very recently, orientations 2 and 4 have begun to assert an influence on science education programs. Orientation 2 can be found in the popular Science-Technology-Society approach (Bybee, 1987; Yager, 1988), and orientation 4 is a part of the environmentally focused global education programs (Bybee, 1984).

My position is that academic rationalism can subsume the other four orientations when epistemological frameworks that focus on both faces of science are included. The key to this subsumption lies in the realization that science has two faces—a testing or justification face and a development or generation face. The theory testing and theory change frameworks presented in this chapter extend the discussion begun in Chapter 4.

AVOIDING FINAL FORM SCIENCE

Theory change is a natural element of all scientific disciplines. It is impossible in our classrooms, given the curricula we are expected to teach, to avoid characterizing science as a dynamic, evolving body of knowledge. The growth of scientific knowledge today is unparalleled when compared with any other time in history. The infusion of science into our culture is gaining importance with each decade that passes, as shown by the list of discoveries in science during this century presented in Figure 4.2 and the examples of theory change listed in Table 4.1. The challenge we face is how to accurately capture the evolving character of scientific knowledge, while at the same time not losing our appreciation for the rationality of science.

Presently, secondary level science textbooks and curricula all too often present changes in scientific knowledge with little regard for the dynamics of what prompted them. What we get is final form science. We are told, for example, that once the accepted view was that the solar system was a geocentric model, that the burning of substances released phlogistron, that marble contained fixed air, and that all rocks were deposited by water. Our textbooks and curriculum guides typically treat these knowledge claims as trivial instances of the past, and after a brief dismissal of these knowledge claims, we are informed that today we firmly believe in the heliocentric model, oxygen, carbon dioxide, and igneous, sedimentary, and metamorphic rocks. How did we arrive at this knowledge? Why did people/scientists accept the older views? What evidence prompted the change from one view to another? Such questions typically remain unanswered and thereby contribute to an inaccurate view of science.

There are three dangers inherent in a final form presentation of science. One danger is the perception that all knowledge claims can be treated equally. That is, new or fringe theories are placed on the same pedestal as frontier or central theories, resulting in an oversimplification of the structure of scientific theories. Scientific theories are complex statements that interact with other theories. Therefore, another danger of final form science is that knowledge claims are taken out of context. That is, theories are reviewed in isolation rather than as elements in a core network of theories that make up a discipline. Such a network, it must be stressed, is arrived at through complex social dynamics of a scientific community.

The final danger is a natural byproduct of the first two. When the structure and role of theories are oversimplified, there is little need to accurately portray the processes of theory change. This is unfortunate, because science is an activity in which the replacement and substitution of knowledge claims, methods, and cognitive aims are ongoing activities. At the level of classroom instruction, these dangers are manifest in the cumulative effect a final form science education can have on developing in learners an incorrect view or interpretation of the nature of science.

One goal of precollege science education programs is to teach students about the tentative nature of scientific-knowledge claims. Science, we stress, is a discipline that continually questions. Evidence that this goal has not been met—that students hold and develop a final form view of science—has been documented in the science education research literature (Rubba, 1977; Welch, Klopfer, Aikenhead, & Robinson, 1981). The prevailing view among learners about the nature of science reflects an authoritarian view of science. Such a view maintains that scientific knowledge is presented as absolute truth and final form.

A significant source of the authoritarian view of science is the belief that the scope of scientific authority is unlimited and beyond reproach. From a final form perspective, if it looks like science and acts like science, then it must be science. Evidence that this has happened can be found in the contemporary debates over equal time for scientific creationism theories and evolution theories and in the propensity of people to accept miracle cures for any number of ailments. Niles Eldredge (1981) places the blame for the acceptance of the "fair play, equal time" argument and of pseudoscientific claims on the authoritarian final form presentation of science in our classrooms. He writes:

> Both systems [creation "science" and evolution] are presented as authoritarian, and here lies the real tragedy of American science education: the public is depressingly willing to see merit in the "fair play, equal time" argument precisely because it views science almost wholly in this authoritarian vein. The public is bombarded with a constant stream of oracular pronouncements of new discoveries, new truth, and medical and technological innovations, but the American education system gives most people no effective choice but to ignore, accept on faith, or reject out of hand each new scientific finding. (p. 15)

The frameworks presented in this chapter further establish an epistemological basis for teaching science which seeks to avoid final form versions of science. This chapter will outline specific criteria teachers can employ to make choices in evaluating the status of scientific-knowledge claims and to teach in light of the two faces of science. In the next section two frameworks are introduced for evaluating theories: Ronald Giere's (1984) test of theories and Robert Root-Bernstein's (1984) four criteria for theory evaluation. The final section of the chapter presents two models of theory change: Larry Laudan's (1984) triadic network for describing changes in cognitive values and Dudley Shapere's (1982) tripartite process of observation. Within each of these four frameworks, science teachers will find useful categories and partitions to guide their selection and sequence of science concepts.

The approach here is pragmatic. Philosophically speaking, the discussions that follow are instrumental. In an attempt to provide a procedural format that teachers and students can employ to assess scientific-knowledge claims, liberties are taken in the selection of philosophic models and frameworks. I recognize fully that each of the models by itself is incomplete as a full and satisfying characterization of the growth of scientific knowledge. However, this does not dismiss the fact that such models provide pragmatic bases for guiding the teaching and learning of science. Furthermore, the argument is made that by adopting a curriculum perspective that evolves from a complex set of scientific processes based on epistemological principles—processes that seek to describe the testing as well as the generation of knowledge claims—a broader set of orientations can be integrated in a K-12 science curriculum.

EVALUATING THEORIES

Being able to rank theories according to the status they hold among a community of scientists suggests that a set of criteria must exist that can be used to assess the adequacy/truthfulness of a theory. Testing scientific theories is difficult. Chapter 4 explained that there are levels of theories and a complex set of variables with which to evaluate theories. Nonetheless, it is possible to put forth a general framework for testing theories.

Testing Theories with Arguments

One useful schema for testing theories has been proposed by Ronald Giere (1984). First, he suggests a scientific theory should be treated as a hypothesis in which a theoretical model is making a claim about the real world. This claim can be called a "theoretical hypothesis." Next, this hypothesis is treated as both a contingent statement—it can be either true or false—and a conclusion of an argument. An argument is a set of premises ($P_1 P_2 \ldots P_n$) that lead to a statement of a conclusion. An argument is analyzed in either one of two ways: (1) by testing the truthfulness of the premises, all of which must be true, or (2) by testing the internal consistency of the set of true premises, in which there can be no contradictions. Putting a theory into argument form, we can test or justify the conclusion of the argument by examining the premises used to develop the argument. In other words, it is possible to format a test of the theory based on the pieces of information that make up the theory.

The basic elements for a test of a theoretical hypothesis are the hypothesis (TH_o), the prediction (*P*), the initial conditions (*IC*), and the background knowledge (*BK*). TH_o is the theory being tested, *P* is the predicted occurrence of the possible state of some real system described by the theory, *IC* are the states of the system before considering this hypothesis—the known facts, if you will—and *BK* represents the existing knowledge claims of science that the hypothesis should not refute, that is, the central ideas of science.

Breaking a theory down into these constituent parts helps us decide where the strengths and weaknesses of the argument lie. As it concerns teacher planning, this strategy enables us to determine the most important content—that which informs us about our scientific theories. Essentially, the strength of the argument is determined by the strength of the theoretical hypothesis in accounting for the prediction while not coming into conflict with either the initial conditions or the background knowledge. Giere's schema for testing a theory, then, is not that dissimilar from the goal-of-science hierarchy introduced in Chapter 4. To challenge the prediction, the conclusion of the argument, we can attack any of the premises. We can attack TH_o, we can attack the *IC* upon which the hypothesis is based, or we can attack the premises of *BK* the hypothesis chooses to include. An example will help clarify the argument-strategy test of a theory.

An interesting episode in the history of science will serve as the basis of our example. At the beginning of the twentieth century Alfred Wegener, a German meteorologist, put forth the theoretical hypothesis that the earth's continents were once together as a single land mass and then drifted apart to the positions we find them in today. In 1926, geologists at the annual meeting of the American Association of Petroleum Geologists soundly rejected this idea of moving continents. For the next 30 years, little attention was paid to drifting continents; in fact, those geologists who continued to profess allegiance to the idea of drifting continents jeopardized their professional standing in colleges and universities. In 1955, however, the idea of drifting continents was reintroduced and ultimately led to our present theory of plate tectonics. This is one of the few instances in which a rejected theory was revived. Looking at the arguments for this theory, using Giere's schema, will help us understand how this change in theory occurred.

TH_o
The continents on the surface of the earth were once joined together as a single land mass.

IC
1. The continents on the opposite shores of the Atlantic can be pieced together like a puzzle.
2. Glacial deposits and plant fossils of the same age are found on the continents of South America, Africa, Australia, and Antarctica.
3. There is not enough water on the planet to make a glacier large enough to extend to these continents given their present positions; nor could the fossils that are found in the rocks transport themselves across the oceans.
4. Geological structures (faults and rock types) of the same age and with similar lithologies and orientations can be found on North and South America and on Africa.

BK
Physical theories hold that the earth is a solid object.

P
The continents on the surface of the earth are moving with respect to one another, or drifting.

P is the conclusion of the argument that was rejected by the geologists in the 1920s. It was rejected for rational reasons (Frankel, 1983), because the background knowledge of geologists, that is, the central theories they held about the structure of the earth, did not provide for other than an earth that was solid through and through. Subsequently, the prediction of drifting continents also refuted the laws of physics, which did not include mechanisms to allow for solid rock to move through solid rock. The premises of the argument were not internally consistent.

Not until a change in *BK* came about could the *P* of drifting continents be advanced. Such a change did eventually come about through the study of earthquakes, or seismic waves. The study of the behavior of seismic waves suggested that the earth was not entirely solid. To explain the patterns of data being collected by seismographs, a partially liquid-core model of the earth had to be adopted. Once the veil of an entirely solid-core earth had been lifted, seismologists also found it prudent to propose that the upper section of the mantle was more accurately described as a plastic substance than as a solid. Shifting the *BK* from a solid-core earth to a partially liquid-core earth was the critical turning point that enabled drift theory to be considered a viable explanation for the varying geomagnetic patterns found in rocks on the British Isles.[1]

The usefulness of this test for science teachers resides in the way it partitions broad theoretical statements into sets of smaller statements—theoretical hypothesis, background knowledge, initial conditions, and a prediction. This partitioning puts teachers in a better position to invoke instructional strategies that lead to meaningful learning by students.

It is a fundamental maxim of current effective education practice that sound instruction seeks ways to partition knowledge claims and then establishes the relationship among the parts. Novak and Gowin (1984) refer to this linking of knowledge claims as meaningful learning. Furthermore, they suggest that as a basis of meaningful learning, learners should develop a sense of the metaknowledge of the topic to be learned. Metaknowledge is the structure of the knowledge to be learned, or the syntax of the network of concepts to be learned. With a familiarity of the metaknowledge, learners are better able to identify for themselves missing elements in the process of meaningful learning. Thus, Novak and Gowin (1984) relate metaknowledge to metalearning. Metaknowledge can be thought of as knowing what to know, and metalearning as learning how to learn. In the case of Giere's framework for testing theories, metaknowledge would represent knowing the components of the argument: TH_o, *IC*, *BK*, *P*. By analyzing the parts, it is possible to both introduce learners to the structure of the knowledge claim to be learned as well as explore strategies for using the structure as a tool for learning.

Thus, if the knowledge claim under investigation is a complex statement like a scientific theory, then effective instructional practice will first establish the parts or structure of the theory. The next step of instruction will establish the relationships that exist between or among the parts of the theory, and the final step will explore the degree to which the theory holds up against prevailing evidence, reason, and social conditions. These three steps, when taken together, help establish the metaknowledge framework of the theory.

The three steps, as originally designed, are part of the questioning strategy advocated in the Pattern of Inquiry Project (Connelley & Finegold, 1977), which follows a pattern of structural, functional, and evaluative questions. Structural questions are what-type questions that help students understand the parts of the inquiry. Functional questions are how-type questions that enable students to understand how one part is related to another part. Evaluative questions are why-type questions that seek to establish the degree of confidence we can have in the knowledge claims being proposed. In brief, this questioning triad seeks to establish the degree of legitimate doubt we can associate with scientific-knowledge claims.

Once subdivided, the theory is capable of being analyzed to establish the degree of legitimate doubt we can attach to it. The task for the teacher is to examine with students (1) the initial conditions the theory must account for or explain, and (2) the background knowledge with which the theory must be consistent. Just like the analysis of any other argument, the premises—the initial conditions and the background knowledge—of an argument must be true and the relationships between the premises must lead logically to the conclusion of the argument; that is, they must not contain contradictions. A theory free of false premises and free of contradictions will carry very little or no degree of legitimate doubt. Central theories and frontier theories typically meet these criteria. On the other hand, a theory that contains false premises and/or contradictions among the premises will have a degree of legitimate doubt attached to it.

The important point is that a theory itself is not replaced, abandoned, or rejected. Rather it is the parts of the theory that are replaced, abandoned, or rejected. In this way we can establish a rational line of argument for the changes that occur in science. Subdividing the theory into parts—a strategy useful for planning and implementing instruction—also is a procedure which requires that we look at those parts. Thus, let us explore further other mechanisms for partitioning theories and theory change processes into parts that can help us in formating instruction.

Four Criteria for Judging Theories

Robert Root-Bernstein (1984), as part of a collective response and critique to the rise in scientific creationism (Montagu, 1984), suggests that scientific theories can be judged by four fundamental criteria: logical criteria, empirical criteria, sociological criteria, and historical criteria. Root-Bernstein's framework extends Giere's logical empirical orientation and also includes two new sets of criteria that focus on forces external to theory change, namely, sociological criteria, and forces internal to theory change, that is, historical criteria. We will explore each of these four sets of criteria in more detail to acquire a sense of the component parts that can enter into our instructional decision making.

Logical Criteria. A theory can be judged by four primary logical criteria (Root-Bernstein, 1984, p. 65). It must be

> (1.a) a simple, unifying idea that postulates nothing unnecessary ("Occam's Razor");
> (1.b) logically consistent internally;
> (1.c) logically falsifiable (i.e., cases must exist in which the theory could be imagined to be invalid);
> (1.d) clearly limited by explicitly stated boundary conditions so that it is clear whether or not any particular data are or are not relevant to the verification or falsification of the theory.

Carefully considering each of these four criteria should make it clear why they are essential to theory evaluation. "Theories must make clear patterns of things and relationships between things" (Root-Bernstein, 1984, p. 65). There are obvious similarities between these four logical criteria and those Giere proposes in his argument scheme for testing theories. Properly construed, a theory will be succinct and straightforward in the explanation it seeks to provide. The theory will not contain statements that contradict other statements it contains. The theory must be testable in terms of both justifying and refuting it. These are Giere's two conditions for a good test.

Good theories provide sound explanations, and sound explanations are based on logically sound arguments. The requirement that a theory be precise in delineating the phenomena it seeks to explain establishes both the domains in which it applies and the domains in which it fails to be relevant. Unbounded theories cannot be falsified and therefore cannot be shown to have any weaknesses. The ability of scientists to identify explanatory weaknesses and logical inconsistencies is basic to the self-correcting nature of the growth of scientific knowledge. "Self-correctability is precisely the characteristic that gives scientific theories their epistemological power: a theory that is incorrect or incomplete can, by attempts to falsify it, reveal its faults or limitations and so be corrected or extended" (Root-Bernstein, 1984, p. 66).

Empirical Criteria. Empirical criteria are critical to the development of a data foundation for the growth of scientific knowledge. We pointed out earlier that the goal-of-science hierarchy began with the analysis of data. The empirical criteria of theory evaluation focus on describing what counts as observable measurable data that can be used to verify tests of theories. The four criteria state (Root-Bernstein, 1984, p. 66) that a theory must

> (2.a) be empirically testable itself or lead to predictions or retrodictions that are testable[2];
> (2.b) actually make *verified* predictions and/or retrodictions;
> (2.c) concern reproducible results;
> (2.d) provide criteria for the interpretation of data as facts, artifacts, anomalies, or as irrelevant.

These four criteria enable us to more clearly understand the rational feedback mechanism in Figure 4.1. Criteria 2.a, 2.b, and 2.c are straightforward and do not require further elaboration. But criterion 2.d is central to the growth of scientific knowledge and a powerful tool for teachers to use in the planning of science instruction.

In criterion 2.d we see once again that all things in science are not equal. Some data counts or is relevant, some is not. I want to focus on data that is initially not construed to be relevant. As new technologies emerge (for example, magnetic compass, microscope, vacuum chambers, cathode ray tubes, electric generators, clocks), new modes of observing and measuring are developed, which, in turn, generate new sets of empirical data. This new data born of new technology typically falls outside the established explanatory boundaries of known theories, but is nonetheless too valid to dismiss.

Such data or empirical facts are referred to as anomalous data and are an important force in changing the explanations of science. When the amount of anomalous data becomes large enough to prompt some scientists to call into question the existing central explanations/theories of science, Thomas Kuhn (1962/1970) suggests science enters a period of revolution, whose purpose is to seek a conceptual change in what data counts as fact, artifact, anomalies, or irrelevant.

The identification of anomalous data by teachers in the course of science instruction provides students with a tangible link to the mechanisms of change in scientific knowledge. As pointed out earlier, such data is frequently linked to advances in technology. Thus, an important aspect of science education is outlining how the processes of observation and measurement have been changed over the centuries to rely less and less on the direct role of the human senses. We will explore this important process of change in the last section of this chapter.

Today, magnetic probes describe the workings of the human body and the characteristics of the earth's mantle. Satellites employing spectra of electromagnetic radiation invisible to the naked eye are used to view the surfaces of planets. Statistical procedures enable investigators to make predictions and retrodictions on subcomponent parts of populations. In each of these examples, the collection and manipulation of the data need not involve the active use of human senses. Computers, state-of-the-art electronics, and theories of the physical universe have made it possible to redefine what it means for something to be empirically testable, verifiable, and reproducible. In turn, these changes have altered how we treat data.

Sociological Criteria. Science does not function in isolation. Similarly, scientists do not practice their profession or conduct their inquiries without influence from the outside. The influence here is directed toward the collective contributions scientists make to each other in the pursuit of answers to scientific problems. When private knowledge becomes public knowledge a ripple is sent across the scientific landscape. Today the forum for the transmission of private knowledge into public knowledge is journal manuscripts and conference proceedings. In nineteenth-century England, the prevalent forum was the meetings of the Royal Society of London. Gentlemen would gather to watch demonstrations and debate the implications of experiments conducted by Robert Boyle and his assistant Robert Hooke or to hear the result of the global explorations of Charles Darwin or Charles Lyell. Earlier still, learned monks and priests would meet to reconcile the implications of Aristotle's views of the universe with those of the church.

The communication of knowledge establishes standards and records what is to count as legitimate knowledge claims, as relevant problems for investigations, and as appropriate methods of conducting an investigation. Consequently, practicing scientists must accept the burden of responsibility to ensure that their explanations have been checked against past as well as existing explanations. It is in this spirit of knowing what else exists that Root-Bernstein (1984, p. 67) offers his four sociological criteria for determining the validity of a theory. A theory must

(3.a) resolve recognized problems, paradoxes, and/or anomalies irresolvable on the basis of preexisting scientific theories;
(3.b) pose a new set of scientific problems upon which scientists may work;
(3.c) posit a "paradigm" or problem-solving model by which these new problems may be expected to be resolved;
(3.d) provide definitions of concepts or operations beneficial to the problem-solving abilities of other scientists.

A scientific theory that ignores the past is always held suspect. Of course, there are often very different opinions about what problems are most worthy of investigation. The history of science is fraught with many examples of periods of consensus and dissensus about the theories, methods, and aims of science. Pointing out examples of these to students is a valuable element of a sound science education. A fuller discussion of this application of philosophy of science follows the discussion of the last of Root-Bernstein's criteria.

Historical Criteria. The last of the four sets of criteria for theory evaluation proposed by Root-Bernstein (1984, pp. 67–68) is historical criteria. Historical criteria ensure that the growth of scientific knowledge has followed a path that clearly establishes correctability. The three historical criteria state that a theory must

(4.a) meet or surpass all of the criteria set by its predecessors or demonstrate that any abandoned criteria are artifactual;

(4.b) be able to accrue the epistemological status acquired by previous theories through their history of testing—or, put another way, be able to explain *all* of the data gathered under previous relevant theories in terms either of facts or artifacts (no anomalies allowed);
(4.c) be consistent with all preexisting ancillary theories that already have established scientific validity.

The knowledge claims of science do not occur within a vacuum. The activities of science and therefore the knowledge claims of science must have meaning within a cognitive and methodological framework of those scientists who are part of the scientific community at large. Thus, we speak of scientists functioning within research programs (Lakatos, 1970) or research traditions (Laudan, L., 1977) and of ideas belonging to certain world views or paradigms (Kuhn, 1962/1970). Of course, the unifying central theory elements of science can change also, but such revolutionary changes in scientific knowledge are a less common occurrence than the frequent steady, cumulative-type change that occurs within a research program or research tradition.

The next philosophical frameworks we shall examine focus on an analysis of what is involved when one central theory of science is replaced by another.

CHARACTERIZING THEORY CHANGE

Historical studies of the growth of scientific knowledge (see Brush, 1988 for a useful bibliography on the history of modern science) have had the effect of altering philosophers' views of what characterizes knowledge growth in science. Prior to the influx of history of science during the 1950s and 1960s, the growth of scientific knowledge was modeled after theories of the physical sciences, which are characterized by mathematical formulations. Those physical science models promoted an accretionary mechanism to describe the growth of knowledge. In the accretionary mechanism, new knowledge is added to old like layers of aluminum foil to an aluminum foil ball.

Scholarly writings in history of science and sociology of science have altered this accretionary view to one that argues for the replacement and substitution of knowledge claims. In terms of our aluminum foil analogy and employing Root-Bernstein's criteria for theories, evaluating anomalous data born from new techniques and technologies for collecting data requires that we sometimes start new balls of aluminum foil. Thus, contemporary principles of the structure of knowledge speak to the importance of establishing a basis for accepting scientific theories as both revisionary and tentative (Suppe, 1977). An adequate philosophy of science, therefore, must understand both how theories are generated and how they are revised, altered, or replaced.

It is this very idea of revision and replacement in scientific theories that has prompted educational researchers (Posner, Strike, Hewson, & Gertzog, 1982) and psychologists (Carey, 1986) to suggest that principles of theory change can be applied to how children learn science. In Chapter 1 we introduced the idea of conceptual-change teaching and conceptual-change learning. It is important to recognize that the application of conceptual change to education is adopted from the ideas of philosophers of science about the revisionary characteristics of the growth of knowledge.

Inasmuch as scientists and children share the common task of seeking ways to expand their knowledge of the world, it is not far fetched to suggest that information gained from an analysis of one group can assist in understanding the other group. For example, this is precisely what Jean Piaget sought to do with his in-depth analysis of the growth of cognitive development in children (Kitchener, 1987). Through a thorough study of the growth of knowledge in children, Piaget hoped he would be in a better position to understand the epistemological foundations of knowledge development. In a sense, the children were the laboratory in which he conducted experiments in epistemology. The results of Piaget's studies have certainly had a dramatic impact on child psychology and cognitive science, but he was not a psychologist by training; he was a genetic epistemologist (Kitchener, 1987).

In this final section of the chapter, we turn our attention to frameworks that can assist us in describing this mechanism of conceptual change. Our task, however, unlike Piaget's, is to apply epistemology to the task of teaching science. The fact that the cognitive processes children employ when learning science happen to have much in common with the epistemological frameworks of science theory development serves as a partial justification for applying epistemological frameworks to teacher decision making.

Arguments that the structure of scientific theories and the structure of learners' cognitive frameworks/schemata have much in common must consider both the structural characteristics and the developmental characteristics of these conceptual frameworks. Structurally, scientific theories and cognitive schemata seem to have much in common. A theory can be thought of as a network of facts, principles, and lawlike statements that are joined together by accepted methodological practices and by common goals or aims of inquiry. A cognitive schema, from this structural perspective, is conceived as a cluster of concepts and propositional statements that are governed by rules of logic and social values that together guide the synthesis of knowledge.

A problem occurs, though, when it is necessary to describe the mechanisms for change in the structure of theories or of conceptual schemata. There has traditionally been among philosophers a clearer sense of what constitutes a logic of justification than of what characterizes a logic of generation. With the recent application of history of science and the emergence of fully developed central theories in biology and geology, philosophers of science have been in a better position to develop guidelines that describe the generation or development of theories.

But we in education have yet to witness a conversion in our curriculum to these new views of science. A negative outcome of the clearer sense of models describing a logic of justifying scientific knowledge has been an exclusion of the dynamics and mechanisms of change in scientific knowledge. In short, the important roles of theory refinement, adjustment, and substitution are excluded from science classes. Included are factors that promote a final form representation, or should I say misrepresentation, of science.

What I want to challenge is the more typical hierarchical straight path growth to a final form view of science. Hierarchical models, like other building-block approaches or accretionary models (such as the aluminum foil ball analogy) of the growth of scientific knowledge, underestimate how important the mechanisms of change are in the growth of scientific knowledge. Simply stated, for now, such models unduly emphasize the successes of science and the final form of scientific knowledge. I do not wish to challenge the position that science has provided increasingly more accurate accounts of the structure of nature. It has. It is difficult to deny the fact that over the past three centuries our scientific understanding of nature has improved.

What I do want to challenge is the one-sided approach that emphasizes the consensus-building mechanisms of change. Although the general trend has been one of success, historians of science have found that this long-term positive trend has many instances of retrograde motion, stasis, and dead-end lines of inquiry. Theory building is more accurately described as an activity involving replacement and substitution than as an activity involving accretion and addition. The conclusions drawn by historians of science suggest the growth of scientific knowledge progresses through periods of consensus and dissensus among practitioners. Furthermore, the changes to theories that occur are not global holistic changes, but rather piecemeal in character. Again, we see a partitioning of the process, and thus yet another opportunity exists for us to base teacher decision making on the parts of an epistemological framework.

When changes in science are described in broad terms—once we believed the earth was at the center of the universe but now we believe the earth revolves around the sun—we are too often left with the impression that changes in scientific knowledge come about quickly and holistically to a community of scientists. We mentioned earlier that Kuhn (1962/1970) referred to the periods of change in which new central theories come to replace old central theories as scientific revolutions. He partitions the activities of a scientific community into two parts: periods of normal science when there is consensus about the central theories of a scientific discipline, and periods of revolutionary science when there is dissensus about the central theories of a discipline. It is through periods of revolutionary science that change and progress come to scientific knowledge.

The problem here is that hierarchical models of conceptual change assume that with changes in commitments to central scientific theories, there are concomitant wholesale changes in all other elements of scientific activity, specifically commitments to investigate methods and aims. Such oversimplifications of the process of change in science eliminates from consideration the essential roles changes in methods and aims play in the process of change. They also underestimate the degree to which replacement and substitution are factors in the growth process of scientific knowledge. One effective way to combat final form versions of science in our classrooms is to carefully examine the process of theory change through an analysis of changes in knowledge, methods, and aims.

The principal focus of our discussion will be two models that describe the processes of theory change: (1) L. Laudan's (1984) triadic network and (2) Shapere's (1982) tripartite process of observation. These models of theory change, as with the models of theory evaluation presented earlier, provide a basis for delineating the parts of the process. With a knowledge of the parts, science teachers have yet another set of guidelines for informing them in their planning and implementation of science lessons and science units.

Triadic Network

Hierarchical models of holistic conceptual change put too much emphasis on the effect of changes in theoretical commitments on changing the methodological and investigative aim commitments of scientists. Such a view of change in scientific theory is faulty for several reasons, according to L. Laudan (1984). First, holistic change models inaccurately describe how scientific consensus and dissensus take place. Laudan argues, with convincing evidence from history of science, that change in science is less holistic and more piecemeal in character. For Laudan, the notion of change is one in which changes to scientific theories, methods, and aims can and do occur separately within distinct and mutually exclusive periods of time. Thus, although there is a change in theory commitment, there need be no instantaneous change in commitments to methods or aims.

The second fault Laudan finds with holistic models of scientific change is the emphasis placed on the role of consensus activities in science. During times when enough anomalous data has accumulated to provide evidence that a change in science is needed, dissensus activities among scientists dominate. Hierarchical models of change that promote final form versions of scientific knowledge tend to smooth over this important replacement and substitution component of scientific change. It is important not only because it more accurately describes the process of change in science, but also because studies in cognitive science clearly demonstrate that learners do not adopt or even consider new explanations until old explanations are shown to be inadequate. Thus, what we know about the dynamics of scientific change we can apply to the classroom. We will explore specific strategies in Chapter 6.

A third problem Laudan finds with holistic models is they frequently overlook the important role aims and methodology have in establishing the anomalous evidence that brings about conceptual change. This point is demonstrated by the two models of theory change described by the goal-of-science hierarchy (see Figure 4.1).

One type of change involves the development of a new central theory, which then causes us to interpret existing data in a new light. Such a change is representative of holistic models where a change in theoretical commitment changes how we see, interpret, and represent data. The second model of central theory change applies when new aims and methods of science give rise to new vistas of inquiry and thus new sets of data. Here is where we might invoke Root-Bernstein's empirical criteria as a learning strategy. When the new data fits into the anomalous category, as opposed to the factual, artifactual, or irrelevant categories, then change is imminent and necessary.

We must ask, therefore, what activities in science spawn anomalous data, that is, data that does not fit with prevailing theories of science but is nonetheless too real to dismiss altogether. Certainly one dominant source of new data is the adoption of new technological devices by scientists. These new devices affect the methods scientists employ. Examples are the impact the microscope, telescope, seismograph, x-ray tube, or computer have had on the form of data scientists have at their disposal. Laudan's piecemeal approach focuses our attention on the important role of the tools, strategies, and technologies of the scientific method in bringing about dissensus activities and, thus, eventuating change in science.

In a similar vein, changes within the aims of scientific inquiry can also provide the impetus needed to spawn new sets of anomalous data. The investigative methods of science determine how we look. The investigative aims of science determine where we look. Consequently, when considering influences on aims, we must recognize the important role social factors have on the activities of scientists and scientific institutions. World War II, as we have already discussed, had an enormous effect on the rate at which new technologies were developed. Our space program is an example of how a scientific institution with political and financial support can produce theory-altering sets of data. Clearly, our knowledge of the earth today, a knowledge forged from data collected by various satellites, is significantly different from the knowledge of the earth we had prior to 1960. One significant example is hurricane tracking and warning.

What is being suggested is that the process of knowledge growth is not governed solely or dominated by commitments to theories. According to Laudan (1984), there are three levels of commitment for scientists: (1) theoretical commitments, (2) methodological commitments, and (3) goal-of-science or aim commitments. In contrast to hierarchical models, which emphasize theory commitments over the other two, none of the three levels in Laudan's model has a privileged status over the other two levels. Rather, Laudan (1984) argues:

> There is a complex process of mutual adjustment and mutual justification going on among all three levels of scientific commitment. . . . The pecking order implicit in the hierarchical approach must give way to a kind of leveling principle that emphasizes the patterns of mutual dependence between these various levels. (p. 62)

Thus, Laudan proposes that knowledge justification in science involves a developmental component, which can be represented as a triadic network where

1 Aims (*A*) justify methodology (*M*) and must harmonize with theories (*T*).
2 *M* justifies *T* and exhibits realizability in *A*.
3 *T* constrain *M* and harmonize with *A*.

This network is shown in Figure 5.1. Whereas hierarchical models (Kuhn, 1970/1962; Lakatos, 1970; Laudan, 1977) argue for wholesale or holistic changes in *M* and *A* whenever changes in *T* occur, the triadic network suggests that changes in scientific knowledge are more piecemeal in character. Hence, changes in *T, M,* and *A* can and do occur separately and within distinct and mutually exclusive periods of time.

According to the triadic network model of change, it is possible for scientists working in a discipline to alter theoretical commitments but still maintain existing methodological and cognitive aim commitments developed from a previous theoretical framework. For example, the theory of evolution has had little, if any, effect on the methods of taxonomic classification (de Queiroz, 1988). Whereas Darwinian evolutionary theory focuses on the common ancestry ordering of organisms, "the established method of classification consists of ordering entities into classes, groups defined by the attributes of their members" (de Queiroz, 1988, p. 238). The advent of evolutionary theory has done little to change the attributes of what makes a reptile *a reptile* or a bird *a bird.* But it has changed how we think of the grouping of organisms, that is, tax-

FIGURE 5.1 The Triadic Network of Justification[a]

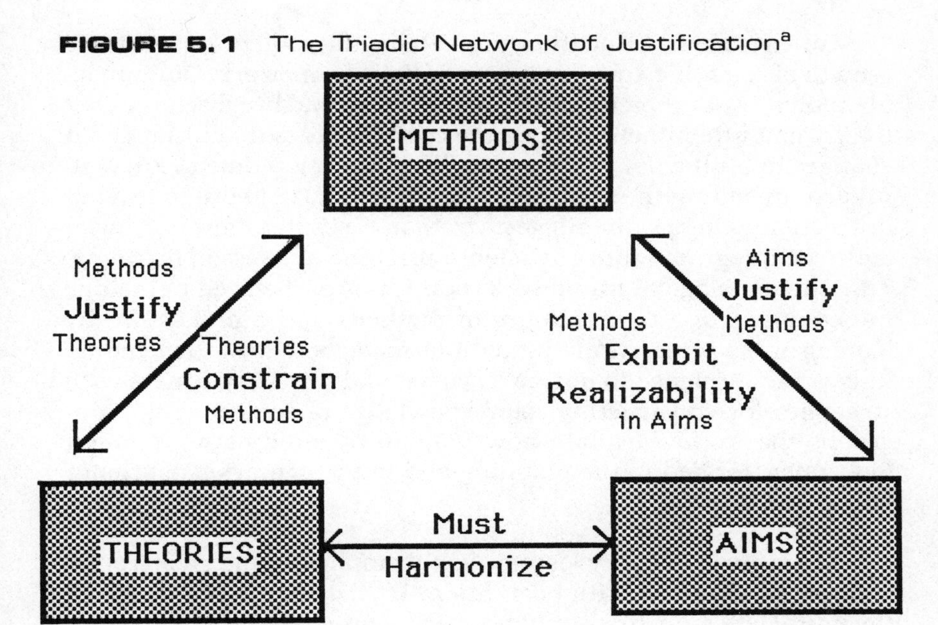

[a] Source: L. Laudan, 1984, p. 63.

onomies based on the natural process of descent, and thus its product, common ancestry, would provide a different way of grouping living things.

The problem with emphasizing commitments to *T* and then T_1, T_2, and so forth, without any discussion of *M*, *A*, or timeframes is it reinforces final form science. Presenting the growth of scientific knowledge as series of theory shifts risks representing the growth of scientific knowledge as a series of whimsical, irrational shifts among scientists. Left aside are the timeframes and the details of the perfectly reasonable and rational events that contributed to the changes in beliefs. Laudan suggests:

> The point, of course, is that a sequence of belief changes which, described at the microlevel, appears to be a perfectly reasonable and rational sequence of events may appear, when represented in broad brushstrokes that drastically compress the temporal dimension, as a fundamental and unintelligible change of world view. (L. Laudan, 1984, p. 78)

We in science education are often guilty of presenting the growth of scientific knowledge in a holistic framework. Our curricula typically paint pictures of change with broad brushstrokes that totally mask the timeframe and the critical events that bring about changes in both what we know and how we have come to know it. In recommending the application of the triadic network to teacher decision making, I am suggesting that a similar "leveling principle" be integrated into our science teaching and lesson planning. This epistemological framework reminds us of the need to include the equally important changes in methods and aims. Again, by looking at the parts, we are better informed about how to meaningfully present science to novice learners, and provide learners with strategies for constructing their knowledge of science. The paragraphs that follow explain how Laudan's triadic network might look when applied to the planning and implementation of science units.

Theories. Theory lessons would emphasize the background knowledge (*BK*) and initial conditions (*IC*), using Giere's terms, of the unit. Here students would need to be provided instruction and learning strategies on how to meaningfully link scientific concepts. The emphasis on concepts recognizes that there is a core content to science. Effective science instruction is stymied, however, if students attach different meanings to the core concepts of the content. Establishing consensus as to core concepts among class members is what conceptual-change teaching is all about. If students' understanding of a concept represents a dissensus among the class, then the teacher needs to implement consensus-forming instructional strategies, such as Giere's and Root-Bernstein's frameworks of theory evaluation. Similarly, if students have a consensus about the meaning of a concept or set of concepts, but the meaning is incorrect, then the teacher needs to invoke dissensus-forming instructional strategies focusing on the development of anomalous evidence, for example, discrepant events.

Methodology. Methods lessons would focus on the acquisition of evidence. A very critical element in establishing anomalous evidence is the changes in our beliefs about what counts as observational evidence. We will explore the issue of observation and evidence in more detail shortly through Shapere's tripartite observation scheme. But here let us again be mindful of the rapid technological advances taking place in this century and how these new tools of inquiry have provided us with new ways of seeing. To expect learners to accept at face value the credibility of scientific data without any clues concerning the process/tools of data collection, data manipulation, and data interpretation is wrong. Such an expectation once again reinforces final form science and presents science as a series of success stories. Left out are the various false starts, dead-end lines of inquiry, adjustments, and misdirections often made by scientists. Methods lessons would impart to students an understanding of how scientists have learned how to learn. In one sense, science is an activity of answering questions that give rise to new questions. How we choose to seek solutions to known questions and to identify new questions is a critical element of the growth of scientific knowledge.

Aims. Aims lessons would focus on what scientists and society believe to be important cognitive goals of science. In other words, these lessons would outline for students what is or was considered most worth knowing. At any given time, there is a set of standards that guide scientists in the selection of important research problems, in the design of methods to solve the problems, and in the establishment of criteria for deciding what counts as a good experiment and, therefore, what constitutes good data or bad data in solving the problem. The formation of standards is a critical element in a proper representation of the growth of scientific knowledge. We have already outlined Root-Bernstein's standards but these are criteria internal to the activities of scientists. There are other important external criteria which influence the scientific enterprise and these criteria fall under the heading *science-technology-society.*

The activities of science, technology, and society do not take place in a vacuum. Social pressures affect the formulation of standards that scientific communities apply to research efforts and research funding. Present day examples of socially relevant scientific problems include global warming of the atmosphere, AIDS, secondary and tertiary recovery of oil reserves, solid-waste disposal, and superconductivity. Historical examples include the development of vaccines for polio, the construction of accurate calendars based on planetary motions, solving the genetic variability of crops, and succeeding at the separation of matter into elements.

Through an examination of the aims of science, we see that there is an applied aspect and a basic aspect of science. We also discover that activities in the arenas of applied science and basic science are equally important to the growth of scientific knowledge. The teaching of how applied science contributes to basic science, and vice versa, is similar. One of the best places to teach students about the interaction of applied and basic science is through examples of how we have learned how to learn about nature. The remainder of the chapter deals with the critically important role observation has in teaching us about nature and about how we have learned to learn about nature. More exactly, we will outline yet another framework for guiding instruction about the interaction of theories and observations.

Tripartite Process of Observation

Here we will turn our attention to the rational development of changes in scientific observation and, therefore, changes in what counts as observational evidence. Up to this point we have focused on strategies to evaluate and represent theories. One other important aspect of understanding the role of theories in science is to examine the influence theories have on the process of scientific observation. Scientific theories represent our best reasoned beliefs about nature and natural phenomena. As scientists have learned how to learn about nature, scientific theories have played an ever larger role in the definition of what constitutes an observation. Increasingly, scientific observations have come to deemphasize the role of the human senses in data collection in favor of instruments and methods based on the fundamental theories of science. The consequence of this shift for our students is that it is becoming increasingly more difficult to explain how we know what we know.

We have learned that what we see is affected by what we know. Thus, although it is possible for two individuals to look at the same source of information, it is nonetheless quite conceivable for them to "see" two very different things. The issue here is the individual's perception of what he or she sees. Hanson (1958) helps us understand this dilemma by distinguishing between two types of seeing: "seeing as" and "seeing that." "Seeing as" observations are carried out without the benefit of prior knowledge. Here observations will focus on the literal description of patterns—a # is seen as two sets of parallel crossing lines. "Seeing that" observations, on the other hand, occur with the benefit of prior knowledge—a # is seen as the symbol for number. Thus, "seeing that" observations have a specific meaning attached to them. When individuals have different knowledge backgrounds, it becomes possible for them to look at the same phenomenon and yet see two different things.

A large measure of what we seek to accomplish in the science classroom is moving students from novice "seeing as" observers or naive "seeing that" observers to informed "seeing that" observers. This is essentially what conceptual-change teaching is all about. Patterns of earthquake epicenters on a world map, direct numerical relationships between an accelerated mass and force, and the geographical range of organisms each imply specific meanings when understood in terms of prevailing scientific theories. For example, recall the observational fact that earthquakes have occurred since the dawn of civilization. Explanations for why earthquakes occur, however, have been numerous and over the last 100 years alone include shifts in barometric pressure, gravitational effects of the moon and sun, isostatic rebound of mountains, plate tectonics, and rising methane gas (Duschl, 1987).

Any attempt to describe the adjustments and refinements that occur in science must provide a schema that outlines changes to the fundamental process of observation. Out of observation emerge the data and evidence that might represent anomalies to what is known. Shapere's (1984) tripartite portrayal of observation enables us to once again partition the role of theories in science and have another framework teachers can employ for making decisions concerning the planning and implementation of science lessons, and students can use as a learning strategy for constructing a knowledge of science.

The ideas expressed here are taken from the collective writings of Shapere (1984). Shapere's philosophy of science argues that not only are there changes to the knowledge claims of science, but there are also changes to our meanings of investigative processes and characterizations of science—what we mean by observation, laws, evidence, and so on. Thus, he further suggests that standards that scientific communities employ are also subject to change. In fact, there is little he would say does not change, including meanings of words. But it is possible to trace the change and then to outline a process he calls the rational feedback mechanism of changes in science. This path is represented schematically in the goal-of-science hierarchy (Figure 4.1) as the adjusting path emerging from the theory level.

In this tradition, Shapere (1982) takes the position that what has come to represent an observational event in science has also undergone a change. Specifically, it is a change where sense-perception evidence has come to take a back seat to theory-laden evidence. Put another way, what seeing meant to scientists in the 1800s is very different from what seeing means to scientists today. The distinction is two very different notions of "seeing that."

Before sketching an example of how theories effect observation—an analysis of how we see shifts in the rotation of the earth—let us examine Shapere's tripartite portrayal of scientific observations. By conceiving of the act of seeing as scientists themselves define seeing and not as philosophers of science define seeing, Shapere (1982) finds it prudent to represent an observation as a three-part process:

1. The release of information by a source
2. The process of transmitting the information
3. The reception of information

Thus, when we think of seeing as a biological process, visible bands of light from the sun reflect off object surfaces. These reflected waves of light travel through our eyes and project an inverted image of the object on the retina of our eyes. If we recognize the pattern of the image, that is, we have a name or label for it, then our receptor—the brain—will be able to act on the information in the appropriate way.

The major premise of Shapere's position is that a data-gathering process that is totally void of sense-perception activities (none of the human senses are involved in the acquisition of data) can still be counted as an observation if certain factors are taken into consideration. First, something is directly observable if information is received by an appropriate receptor and is transmitted without interference to the receptor from the source of the information, namely, the object being observed. Second, observation is composed of the three steps listed above. This tripartite division of observation is significant because within each of the three component observation processes—source, transmission, receptor—scientists have built up theories that relate directly to each phase as a part of observation. Thus, in our scientific knowledge there are theories of the source, theories of transmission, and theories of the receptor. When we consider how this tripartite model applies to any single observational episode in science, for example, seeing the core of the sun, observing the movement of crustal plates, or seeing the structure of recombinant DNA, it is clear that the nature of the evidence that supports our best reasoned beliefs—our theories—is a product of the cognitive knowledge, the methods, and the aims of our investigations. We have learned how to learn about nature.

Let us consider how scientists are using distant quasars and satellite lasers to observe variation in the rotation of the earth (Carter et al., 1984). The quasars and the satellites both emit electromagnetic waves and represent the source of information. With respect to the use of quasars, an important piece of background knowledge about the source of the information (*theory of the source*) is that our present knowledge of the universe treats these distant stars as fixed points relative to the motion of the earth. Therefore, it is possible to determine the distance to the point and treat it as a marker. *Theory-of-transmission* information proposes that electromagnetic radiation can travel uninterrupted from these distant sources to the earth. Our confidence in this theory-of-transmission information is supported by the transmission of millions of lines of communication signals in our atmosphere. (Ever try explaining to a child why the radio signals do not all play at the same time?)

The background knowledge (*theory of the receptor*) of the appropriate receptor for information transmitted by distant quasars is quite diverse and indicative of to what extent scientific reasoning and technology have become a part of scientific observation. The receptor of the information is a global network of radio telescopes linked to sophisticated computers. The entire technique is called very long baseline interferometry, or VLBI for short. The reliability and validity of the instrumentation and of the theories involved are firmly established among the scientific community.

> The improved series of earth-rotation measurements presented here represent some of the early results of a concerted international effort to apply advanced technology to the study of geodynamics. In September 1983 Project MERIT (monitor earth rotation and intercompare techniques of *observation* and analysis) . . . launched a 14-month *observing* campaign to study the earth's rotation. The VLBI, SLR (satellite laser ranging), and LLR (lunar laser ranging) *observations* collected under Project MERIT are expected to yield the most detailed and accurate records of variations in the earth's rotation ever compiled. . . . The improved *observational* capabilities may well allow questions that have dominated the study of geodynamics for more than a century to

> be answered within the next decade. (Carter et al., 1984, p. 961, emphasis mine)

Advances in physical knowledge, in mathematical procedures, in computer science, and in scientific instrumentation have generated a rational feedback mechanism that has changed our concept of observation. "Thus, we say that what we have learned about the way things are has led to an extension, through a natural generalization, of what it is to make an observation" (Shapere, 1984, p. 345). Essentially, we have learned how to learn, and there is a developmental character to this process.

The theories scientists employ are an integral part of understanding what counts as an observation and what counts as evidence. When science education programs focus on final form science, they avoid articulating the important theories of source, theories of transmission, and theories of receptors that contributed to the development of that final form knowledge. An examination and articulation of source, transmission, and receptor theories help students understand the rational evolution of scientific criteria, which in turn helps them better appreciate how scientists have come to think the way they do and to use the methods they do. It also provides students strategies they can use to learn science.

The knowledge claims and investigative methods of science are quite remote to the average citizen. W and Z particles viewed by supercolliders, black holes seen by radio telescopes, million-year-old rocks measured by potassium/argon dating methods, models of biological and biochemical evolution derived from statistical probabilities, gene splicing with enzymes acting as biological scissors—all of these are examples of the claims and methods of science that make up our contemporary understanding of nature. This is strange, wild stuff when only the final form of knowledge is presented without any information about how we have come to arrive at this knowledge.

Through an examination of the development of scientific theories and scientific explanations, students can learn how scientists have not only acquired knowledge about nature but also learned how to observe nature. Modern scientific knowledge and technology are meaningless when treated in isolation; teachers *must* include the background addressing how science arrived at a particular knowledge claim. Through an examination of the rational evolution of scientific accomplishments, an analysis of the development of knowledge and of the methods and criteria for seeking and acquiring the knowledge can take place.

NOTES

[1] An excellent account of the development of the acceptance of drifting continents can be found in Takeuchi, Uyeda, and Kanamori (1970).

[2] Predictions and retrodictions are logically equivalent. The former make statements of explanation about future events, while the latter make statements of explanation about past events. Historical explanations in the geological sciences often employ retrodictions. The classic example is the case of the evolution of the horse hoof described in Chapter 4.

Chapter 6

Learning as a Restructuring Process

The approach to science education presented in the previous chapters has emphasized considerations for the structure of the subject matter. It is not a novel approach. Others, most notably Bruner (1960) and Schwab (1962), have argued that knowledge of the structure of a discipline can be used by curriculum writers and teachers to format instruction. What is novel to the approach herein is the emphasis given to the central role of theories in the structure of scientific knowledge. In order to say we have developed a knowledge of science, we must be able to say we have an understanding of the function, structure, and generation of scientific theories.

Emphasizing the structure of knowledge does not suggest that other factors are not important or even that they are less important in education than epistemological factors. In fact, Schwab (1969) himself later identified four fundamental aspects of all teaching situations he called commonplaces of education: students, teacher, subject matter, and milieu. Each, Schwab argued, ought to be taken into consideration in designing or implementing instruction. The position taken here is that considerations for the structure of scientific subject matter have been a neglected aspect of science education decision making in general—and when considerations for scientific theories are included, the neglect is even greater.

One result of this neglect is the development of a science curriculum that has become philosophically invalid (Hodson, 1988). Such neglect, however, need not persist. As we have learned in previous chapters, the dynamics of knowledge growth in science are linked to the dynamics of theory development, and theory development is best described as a progression of shifting conceptual and methodological frameworks. At a minimum, then, it becomes possible to develop a proper role for theory in our science classrooms. But there is more that can be gained here. With a proper role for theory in our lessons, Hodson (1988) accurately points out that it is possible to have "harmony between the philosophical and psychological principles underpinning the curriculum" (p. 28).

The view that scientific knowledge and theories are the product of a series of replacement and substitution activities has direct application to contemporary views of how children learn. In Chapter 5, we made comparisons between scientific theories and cognitive schemata. Structurally, theories and schemata seem to have much in common. The view of learners emerging today is that they, too, continuously revise and replace their theories as they acquire experience and understanding. Simply stated, "cognitive development involves conceptual change" (Carey, 1986, p. 1127).

In this chapter, we will first explore in more detail what we now know about the process of learning. Next we will outline how teachers' knowledge of scientific theories can assist students in developing learning strategies that foster the revision and construction of their personal scientific knowledge. The study and analysis of scientific theories can provide an instructional context that facilitates both the teaching and learning of science.

LEARNING: A CONSTRUCTIVE ACTIVITY

Behavioral psychological models have dictated and continue to influence much of what it is we do in education. Behaviorally written learning objectives, positive and negative reinforcement, and behavior modification strategies are some of the applications we recognize in our classrooms. It is not possible in these pages to adequately review the development of schools of psychology that inform contemporary educational practice. It is important to recognize, though, that along with the changes in science education and philosophy of science from 1950 to the present, developments were also taking place in brain research and artificial intelligence, which contributed to a new emerging field of cognitive science. Resnick (1983) lists three relevant conclusions that emerge from cognitive science research:

1. The capacity to learn at any given time is limited.
2. An individual constructs meanings and this ability to construct meanings is influenced by prior knowledge.
3. Individuals employ and invent rule-directed procedural de-

vices to expand their capacity to learn and to construct meanings.

These three discoveries about how we learn have powerful implications for how we conceive of the process of education. In general, these new views of learning suggest that meaningful learning is an active process of linking new knowledge with old knowledge, or the unfamiliar with the familiar. The new views of human cognition maintain that students need to be active agents in the learning process. We will consider each of these three conclusions more closely and then relate each of them to the teaching of science.

The capacity-to-learn conclusion informs us that the amount of information we can process is limited. The limit is defined as seven chunks of information plus or minus two. What constitutes a chunk of information, however, is a matter that can be altered through the process of meaningfully relating one chunk of information to others. For example, consider the following sequences of letters: MVEMJSUNP, MTWTFSS, HOMES. If you were asked to learn each of these sequences of letters, you would most likely have the easiest time with the sequence HOMES since it meaningfully comes together as a single chunk of information, a word. The other two sequences have nine chunks and seven chunks, respectively. These two larger sequences can also be chunked together to make a single bit of information after a meaningful relationship between them is established. The first sequence can be represented as the order of planets in the solar system, while the second can be learned by recalling the seven days of the weeks.

Bransford and Stein (1984) have written a delightful primer on learning strategies that evolved from their research and teaching of human cognition. Their book is filled with useful cognitive strategies to enhance one's ability to learn and remember new information and to solve and understand complex relationships. The key is to employ rule-driven procedures that meaningfully group information into more easily accessed larger chunks of information. This is precisely what Resnick's third conclusion tells us.

Studies comparing experts with novices (physics professors with first-year physics students) have found that experts develop procedures (heuristics, rules, categories) that enable them to consolidate information into larger and larger chunks. We can also think of these procedures as a way that experts develop "seeing that" strategies. What distinguishes an expert from a novice in any field is the ability to recognize that a certain pattern of information represents a basis for following a particular strategy or drawing a particular conclusion. In this way the capacity to learn is enhanced.

Given the sequence MVEMJSUNP you could now produce the order of the planets, but the task is made even simpler when the order of letters is expanded into a meaningful sentence: "My Very Elegant Mother Just Sat Upon Nine Pins." Recalling a single sentence now replaces the cognitive task of remembering nine separate words in the appropriate sequence. Thus, the sentence is a chunk of information, just as Every Good Boy Deserves Favors is a chunk for the notes of a musical G clef staff. Similarly, HOMES can be viewed as an acronym for the five Great Lakes—Huron, Ontario, Michigan, Erie, Superior—and EVEN in Avogadro's even hypothesis represents Equal Volume, Equal Number (of atoms).

Learning is enhanced, then, Resnick's second conclusion informs us, when new information is related meaningfully to a learner's prior knowledge. That is precisely the reason effective teachers take the time and make the effort to link concepts from separate lessons into meaningful wholes. The key, it seems, to successful science teaching is to develop in learners strategies that assist them in the construction of larger and larger chunks of information. This is one way the epistemological frameworks in Chapters 4 and 5 can be used.

The idea, however, that learning involves the linkage of information assumes that the prior knowledge is worth linking to. The cognitive science research indicates that if a learner harbors an incorrect or a faulty concept about some phenomenon (for example, the size of an object is directly related to how fast it moves in free fall), then any attempts to link new information to that faulty concept or to build a meaning network from that point of view are doomed to fail.

The identification and alteration of learners' naive views of how the world works are the centerpieces of conceptual-change teaching. It is not enough any more, research informs us, to assume that children's views of science will be replaced in time by the more sophisticated views of scientists. We are learning that teachers must carefully select and sequence the new information that is taught to students. At a minimum, it seems, teachers of science must provide instruction that

1. Identifies learners' initial understanding of the concepts
2. Links concepts together
3. Provides strategies to guide these linkages.

The epistemological frameworks outlined in Chapters 4 and 5 can now be viewed as tools teachers can use to format instructional strategies in ways that are consistent with new views about learning. Cognitive science has made considerable progress in describing the nature of individual knowledge (the three conclusions above); and these descriptions appear to be similar to those offered by philosophers of science for scientific theory. Studies in both domains distinguish two kinds of change in knowledge structures—one frequent, steady, cumulative type of change called weak restructuring, and the other a rarer, less continuous, and noncumulative type called radical restructuring (Vosniadou & Brewer, 1987).

A review of the literature on cognitive science reveals that researchers and scholars are willing to join psychological concepts focusing on schema theory with philosophical concepts addressing theory development (Carey, 1985a; 1985b, 1986; Rissland, 1985). As indicated in Chapter 5, the process of theory development by scientists has often been compared with the development and acquisition of an individual's knowledge of the world (Krupa, Selman, & Jaquette, 1985; Piaget, 1970). Certainly at an intuitive level it is possible to appreciate how each of these domains is concerned with the growth of knowledge. Contemporary developments in history and philosophy of science and cognitive science suggest the growth of knowledge both in a field of study (theory development) and in individuals (schema building) have much in common and can be described by a common language.

Educational researchers interested in learning more about children's views of science have discovered that the knowledge claims employed by learners are not often well grounded by sound rules or relevant associations of concepts (Osborne & Freyberg, 1985). Research in students' alternative frameworks, for example, has found that young learners frequently have knowledge frameworks that, when taken together, are inconsistent. Such results have led Carey (1985a) to argue that the task of understanding the cognitive development of children can no longer be divorced from the task of understanding the processes by which learners change schemata.

Our attention in the remainder of this chapter will focus on the application of epistemological frameworks for describing, evaluating, and modifying the structure of scientific theories to the teaching and learning of science. More exactly, we will review the various epistemological frameworks and consider how each can become a learning strategy to assist students in the activity of constructing and revising their knowledge of science.

CONSTRUCTING SCIENTIFIC KNOWLEDGE

We have discussed earlier in the book how it is possible to partition scientific knowledge into categories for testing and evaluating scientific theories. These categories, or parts, can be used by teachers as instructional design frameworks for linking together science concepts. More important, they also provide a procedural, or rule-driven, mechanism for expanding students' capacity to learn and construct meanings.

The importance of the partitioning of knowledge in science education is that careful attention is given to the structure of the subject matter. Procedural knowledge is knowledge of form and function. It is knowledge of the heuristics and strategies that are used to apply, analyze, synthesize, and evaluate the declarative knowledge of a particular discipline: the facts, principles, theories, and explanations of the discipline. Procedural knowledge is knowing how to recognize problems and strategies for solving them.

Here we will review how the structural characteristics of scientific theories introduced in Chapters 4 and 5 are compatible with cognitive models of learning. Furthermore, we will demonstrate how the partitionings of those scientific theories better enable us to meet the challenge of identifying and formating the procedural knowledge of science.

We have introduced six epistemological models that attempt to explain or characterize knowledge growth in science.

Goal-of-science hierarchy
Levels of theories
Argument pattern for testing theories
Four criteria for theory evaluation
Triadic network
Tripartite process of observation

A common denominator of each of the models is the focus on theories. While it is generally agreed among philosophers that scientific theories represent our best reasoned beliefs about nature and natural phenomena, and are therefore centrally important to the growth of scientific knowledge, teachers and secondary science curricula do not maintain a similar consideration for scientific theories in our classrooms (Benson, 1989; Duschl & Wright, 1989; Hodson, 1988). The six models described herein serve a dual purpose as

1. Procedural knowledge frameworks to guide teaching and learning.
2. Means of providing a more accurate and valid account of the scientific enterprise.

Let us review each of the six models and consider more carefully how each serves as a procedural-knowledge framework.

The goal-of-science hierarchy places theories within a general scheme that seeks to establish explanations and understandings of the natural world.

Scientific understanding
Scientific explanations
Scientific theories
Scientific patterns/laws
Data/facts

This shows the pecking order, if you will, of the status of knowledge claims in science. It also shows the need to subject evidence—data and facts—to analyses that seek to ascertain patterns among them. We have justified for learners why the need to establish patterns and relationships (for example, directly proportional and inverse relationships between variables) is critical. Here we see that the development of such patterns and lawlike statements of science (F = ma; PE = mgh; Mohs hardness scale) do not represent endpoints in scientific inquiry. Are they critical steps in the growth of scientific knowledge? Yes. Are they the goal of scientific activity? No.

The existence of a pattern or lawlike relationship must be explained, and our explanations are our theories. Invoking the goal-of-science hierarchy, therefore, requires learners to first identify and then to relate explanations, theories, patterns, and data. The path here is a two-way street. Given data, what patterns emerge and what theories explain? Given an explanation, what theories, patterns, and data justify such a conclusion? When the context of learning science centers on the evaluation and development of scientific theories, a broader network of relevant concepts emerges.

Consequently, when teachers or textbooks introduce data, facts, or make explanatory claims, students should be motivated to seek answers to the above questions. In science, each knowledge claim is often supported by other knowledge claims. In other words, there is a meaningful network among the knowledge claims of science. Theories explain facts, data, and evidence. Data gives rise to patterns, which need to be explained. When and where meaningful relationships cannot be found, then perhaps one of the theory evaluation/testing frameworks needs to be used to make sense of the knowledge claims.

Our second framework is the *levels of theories.*

Central theories
Frontier theories
Fringe theories

It is not sufficient to place into one broad category all explanatory statements of science. Scientific theories change and evolve and have a developmental history. If we could limit our analyses and study of scientific theories to final form versions of theories, then perhaps we might be able to justify a singular classification for theories. Such was the goal of philosophers of science (logical positivists) during the first half of this century. For them, there was one correct form of physical science theories and that form should be applied to all other disciplines. But, as we have already seen, this view of science is inadequate, because all things in science change, even the content and form of major theories in a discipline.

The three levels of theories provide an educational context that can be used either to distinguish the developmental steps of a single theoretical claim (such as the theory of evolution) or to compare and contrast different theoretical claims important to a field of science at a particular time (for example, Avogadro's even hypothesis, Boyle's law, Charles' law, and the kinetic molecular theory). With the levels-of-theories and goal-of-science hierarchy frameworks, we have a set of labels that makes it possible to teach about differences in explanatory statements and to reinforce the important point that all things in science are *not* equal.

As pointed out earlier, the misapplication of the term *theory* in our classrooms and science textbooks is a manifestation of the misapplication for the term in our language. We have, if you will, two common meanings for the term. One is scientists' use of the term to imply an explanatory statement. The second is nonscientists' use of the term to imply an idea. When the two usages clash, as they have come to do in science textbooks that place disclaimers of "only a theory" or "scientists think" next to theoretical statements concerning evolution, age of the earth, expansion of the universe, movement of crustal plates, or quarks, a significant problem arises. The problem is that students and teachers are quite willing to judge competing statements of explanation on equal terms. The worst case scenario is when scientific theories are equated to scientific facts.

The equal-time controversy between teaching evolution and creation science is a prime example of the misapplication of the term *theory.* The theory of evolution is a powerful explanatory statement that we placed earlier within the transition zone between frontier and central theories. Creation science, the competing explanation, is a fringe theory. The theory of evolution was once a fringe theory, too, but its developmental history—Darwin's natural selection, the great synthesis with Mendelian genetics, the contemporary co-evolution studies, and microbiological evidence—has established without a doubt the centrally important role this theory has among members of the scientific community. When Giere's test and Root-Bernstein's criteria for evaluating theories are applied, the explanatory differences between the two theories emerge.

As Radner and Radner (1982) correctly conclude, only by taking a walk along the fringe of science do we learn what science is and, more important, what science is not. All new explanations begin as private knowledge to either a single scientist or a community of scientists. When the explanation is shared with others, it becomes public knowledge. The entry point of this newly established public knowledge is the fringe level of theories. Some fringe theories give rise to revolutionary thinking in science (for example, evolution, plate tectonics, gravity, electromagnetism), while other fringe theories represent crank ideas (such as parapsychology, Von Daniken's theory that intelligent life visited ancient civilizations, astrology, quartz power, scientific creationism).

Attacks on the credibility of science are becoming more frequent these days. Borderline cases of science are often brought to the classroom by students in the form of newspaper articles, tabloid journalism, television, movies, and popular magazines. Students are typically excited by these novel ideas, and the experienced teacher knows how to capture this enthusiasm. Textbooks do not include many fringe theories; textbooks provide accounts of the central and frontier theories of science. It is important, then, that we have science educators who can address and analyze the fringe cases rather than avoiding them. First, it is important to establish where and how new theories emerge. Second, the ability to discuss what should or should not be considered science for rational reasons is a desirable trait for science teachers to possess. At a minimum, let us distinguish among the levels of theories operating within scientific disciplines.

The levels-of-theories labels, as we have seen above, open the doors for applying Giere's theory testing argument pattern and Root-Bernstein's four criteria for theory evaluation as procedural knowledge frameworks. Consider the general framework of each and how it could be employed as a learning strategy by students.

Theory Testing	*Criteria in Theory Evaluation*
Theoretical hypothesis	Logical criteria
Background knowledge	Empirical criteria
Initial conditions	Historial criteria
Predicted occurrence	Sociological criteria

In the same way that the goal-of-science hierarchy provides a path for learners to follow, the theory testing argument pattern and theory evaluation criteria also provide a context for learning science. Establishing among learners the need to seek out the background knowledge (*BK*) and the initial conditions (*IC*) of a theoretical hypothesis (TH_o) helps identify core concepts and establish meaningful relationships among concepts. Similarly, invoking the logical, empirical, historical, and sociological criteria establishes metaknowledge and metalearning guidelines in science. In short, students are again provided strategies to organize and make sense out of the myriad concepts of science.

In Chapter 5, we discussed at length the importance of anomalous data in the growth of scientific knowledge. It is no less important in the growth of scientific knowledge among students. Students should be encouraged to speak up when new data, patterns, and evidence present themselves as anomalous information. Students should be encouraged to debate conflicts of meaning with the teacher, with the text, and among themselves. Similarly, teachers should be able to identify facts as having once been anomalous data. This is how we can invoke consensus and dissensus in our classrooms.

The four criteria of theory evaluation provide subject-motivated grounds—metaknowledge—for guiding instruction and learning. Contemporary approaches to metaknowledge/metalearning teaching and learning strategies talk about the need to teach to the misconceptions held by learners (Anderson & Smith, 1986; Novak & Gowin, 1984; Osborne & Wittrock, 1983). That is, we can no longer assume that merely teaching the scientific view is a sufficient condition for teaching science. The research of student learning strongly indicates that learners' conceptions or views of scientific concepts, based on their personal experiences and interactions with nature, are strong and resistant to change (Gunstone, White, & Fensham, 1988). Moving learners from their naive views of science concepts to scientists' views of science concepts requires a rigorous and time-consuming instructional plan. At a minimum, an effective instructional plan must include episodes that address both declarative knowledge—what we know—and procedural knowledge—how we know.

Clarification of concept meaning and the development of correct meanings among concepts is the domain of declarative knowledge. It is an essential component of science classes. Each discipline of science taught at the secondary level—life, earth, and physical sciences, biology, chemistry, and physics—has a set of core concepts students must learn and know as scientists know them. To learn and know as a scientist means much more than just rote memorization of terms so that one is able to talk like a scientist. To know something in science is to understand the relationships between and among concepts. Thus, as we have already discussed, knowledge has a procedural component. Procedural knowledge is knowledge of rules, path solutions, heuristics, and the like, which guide us in the construction of meaningful relationships among concepts.

The epistemological frameworks presented in Chapter 5, theory testing, four criteria for theory evaluation, triadic network, and tripartite observation, were presented more as procedural-knowledge frameworks than as declarative-knowledge frameworks. Up to this point we have been stressing how the frameworks can be used to guide teacher decision making. More exactly, it has been a goal of the book to outline how the epistemological frameworks can be used to select and sequence science instruction units. The frameworks establish a procedural outline teachers can follow to format instruction. In the same way, the frameworks can provide students with a procedural outline to learn. Therefore, it is equally important to provide procedural-knowledge frameworks students can use to identify for themselves when there is a problem of establishing meaningful relationships.

A fundamental problem with education, it seems, is that we tend to neglect teaching students about the little tricks we use to learn and to know (Linn, 1987). This book began by recognizing that teachers do not have enough time in the classroom. Many of the problems in our science classes are imposed by curricula that attempt to include too much in one course. Our secondary school science courses are by design survey courses of the scientific landscape. When teachers get caught up in the race to *cover* the material, we teach the declarative without the procedural.

Too often we assume that learners will naturally develop procedures that facilitate learning. This is not the case. Teachers need to include procedures to help students actively and meaningfully link knowledge claims. The epistemological frameworks represent such procedures. When students read science textbooks, they should be encouraged to identify what background knowledge and initial conditions are needed for the knowledge claims they are entertaining. Some of these claims may come from the chapter being read, but others may come from other chapters and even other courses. What science teacher would not want to encourage students to develop learning strategies and metaknowledge skills that seek to link prior knowledge with new knowledge.

Not only do the theory testing frameworks of Giere and Root-Bernstein encourage such links, but they also provide labels for guiding the learning strategy. Thus,

Students can be taught to ask:	*And thereby learn to:*
What are the background knowledge claims of the theory?	Identify the core concepts.
What are the empirical criteria of the theory?	Consider the methods of data gathering being employed.
How does this theory fit with other theories in the discipline and in other disciplines?	Develop a historical developmental perspective of knowledge claims.

The long-term task is to develop learners' ability to learn for themselves. If learning is properly construed by cognitive scientists as an activity of constructing knowledge, then students need to be mentally active. Science as inquiry can no longer be interpreted by teachers as a hands-on approach to science. Today, science as inquiry must also mean a minds-on approach to science. The epistemological frameworks described herein provide examples of minds-on procedural knowledge guidelines students can use to assist them in the personal activity of constructing, revising, and altering scientific knowledge.

These procedural-knowledge frameworks serve to remind teachers of the complex structure of the discipline. It is important that we provide both accurate and complete versions of science to learners. The triadic network and tripartite observation frameworks assist teachers and students with strategies for understanding how science changes. We discussed earlier the important part anomalous data has in bringing about changes in scientific knowledge. The triadic network reminds us that such change is not haphazard. Therefore, we need to balance our science instruction among commitments to knowledge, commitments to method, and commitments to aim. Changes in knowledge are often brought about by changes in method or changes in aim. Providing complete versions of science means being able to explain or justify the impetus for change.

The tripartite process of observation reminds us how far science has developed as an institution that is governed by its own rules and standards. In the process of learning about nature, we have also learned how to learn about nature. A consequence is that what counts as an observation, and thus as evidence, is today described by a complex set of background theories and initial conditions. It simply is not sufficient to tell students our final form versions of what we know. Students must have some insight into how scientists "see." We have stressed again and again the importance of helping students understand how scientists have arrived at a particular stage of scientific knowledge. The path taken is not smooth. Nor is the final destination a simple concept. Theories of the source, transmission, and receptors again detail the network of concepts that, when linked, provide meaning to what we know in science and how we have come to acquire such knowledge.

The growth of scientific knowledge follows paths that contain errors, misdirections, and false starts. There are internal pressures scientists impose on themselves, such as competing theoretical commitments, and there are external pressures with which scientists must cope, for example, federal funding programs. The destinations of the paths taken by scientists are the explanations and theories they invoke to account for the patterns of data and the relationships among types of data that they and other scientists discover, generate, and justify.

The last two chapters of the book present specific examples for using the epistemological frameworks outlined in this and previous chapters. A few summary comments are warranted, before we engage in specifics. First, the common denominator of each of the frameworks is scientific theory. Scientific theories represent scientists' best reasoned beliefs for explaining nature and natural phenomena. It has been argued that making scientific theories the fundamental context of science instruction necessitates outlining in some detail the network of concepts, methods, and aims that meaningfully link together to make up the theory.

Second, the frameworks highlight the equally important need to address both the context of justifying scientific knowledge and the context of generating scientific knowledge in our classes. It is important for our students to know about and to appreciate the need for testing scientific knowledge. How else could we feel confident in our explanations? Therefore, we have a set of guidelines that help us establish a logic for testing.

Unfortunately, scientific knowledge does not achieve a permanent state; instead, it changes. Throughout we have used the term growth to describe the change of scientific knowledge. Are we to assume, then, that our previous tests have been incorrect or just plain bad? We could, but this would be an unfair judgment of our scientific-minded ancestors. The alternative is to examine the rational evolution of scientific knowledge and realize that changes can and do occur in knowledge claims, methods, and aims of science. The character of the growth of scientific knowledge is not strictly cumulative and hierarchical, leading to final forms. Rather, the study of history of science informs us that the growth of scientific knowledge is more accurately portrayed by the terms revision, modification, substitution, and alteration. Understanding the bases for changes in science—in the meanings of terms, the methods used, and the goals of inquiry—requires that we formulate a set of guidelines that describe this ongoing process of discovery and development. Thus, we have also outlined and argued for the existence of a logic of discovery in science.

We spoke earlier of these two contexts as the two faces of science. The epistemological frameworks adopted here were selected as teacher decision-making frameworks and student learning-strategy frameworks because collectively the frameworks address both faces of science. With these scientific theory frameworks, we have a powerful mechanism for selecting and sequencing science instruction. Let us now turn our attention to specific examples of applying the frameworks to instructional design.

Chapter 7

Applying the Growth of Knowledge Frameworks —Chemistry and Physics

Examples of how to apply the specific frameworks to instructional design and pedagogical decision making have been interspersed throughout the previous six chapters. Let us now turn our attention to four curriculum units that illustrate how the epistemological frameworks can be used to select and sequence science instruction. These units are one in each of the principal secondary science disciplines—chemistry, physics, life science, and earth science. Chemistry and physics are outlined in this chapter, while life science and earth science are presented in Chapter 8. The goal is to outline how each of the decision-making frameworks might be used in discipline-specific curriculum topics addressing theory change and the growth of scientific knowledge.

The illustrative curricula represent four storylines or themes of science that are relevant to our present day understanding of nature and natural phenomena.[1] Within each theme, the instructional goal is to establish both what we know and how we have come to know it. The four themes are:

1. The development of the particulate view of matter. How did we come to see gases, liquids, and solids as being composed of elements, molecules, and atoms?
2. The development of the heliocentric model of the solar system and the views that motion is governed by gravity. What observations and observational patterns represented anomalous data to the geocentric model? How did we come to separate motion into components and thereby distinguish constant acceleration from constant speed?
3. The development of the view that the earth and life on earth

have an elaborate history that far exceeds the presence of humans. What is the evidence and the reasoning that allow us to date the age of earth as hundreds of thousands of years old? To invoke a model of life on earth governed by a theory of evolution?

4. The development of the model of plate tectonics. What data contributed to the view that the earth is composed of plates moving relative to one another?

It is not the purpose of these curriculum unit storylines to provide full historical accounts of the development of science in each of these four domains. Rather, their purpose is to demonstrate how teachers can address the dilemmas of deciding what to teach. The historical approach provides some insights into what content to select and how to sequence it. What is being attempted here is to place such content into a pragmatic schema of procedural knowledge that allows us to make informed decisions about selecting and sequencing content. By identifying the most important content for learners and by identifying the rational development of scientific knowledge claims, we become more effective instructional designers and pedagogical decision makers. An important task is to meaningfully present the bases for what we know in science and how we have come to know it; that is, to avoid Kilborn's (1980) epistemological flatness. In this way we can begin to assess the degree of legitimate doubt that can be attached to scientific-knowledge claims.

The liberties taken in the presentation of science herein are taken with a keen awareness of what constitutes the task environment of the secondary level science teacher. The purpose here is one of application: to demonstrate how the epistemological frameworks might be used. In no way should the reader construe the accounts of science presented in this and the next chapter as complete accounts of the history of science. The appendix at the end of Chapter 3 lists references to such accounts.

We will refer to the examples respectively as matter, motion, acquiring a history of life on earth, and moving plates. The fundamental goals of applying the epistemological frameworks to these four scientific domains is to describe a pragmatic strategy—a procedural knowledge guideline—for teaching and learning about scientific theories. To facilitate the presentations, the codes for each framework shown in Figure 7.1 are used.

FIGURE 7.1 Codes for Frameworks

TESTING THEORIES

TH_o	Theoretical hypothesis
BK	Background knowledge
IC	Initial conditions
P	Predictions

FOUR CRITERIA FOR THEORY EVALUATION

L	Logical criteria
E	Empirical criteria
H	Historical criteria
S	Sociological criteria
A_mD	Anomalous data

LEVELS OF THEORIES

C	Center theory
Ft	Frontier theory
Fr	Fringe theory
N_f	Novel fact

THEORY CHANGE—TRIADIC NETWORK

T_c	Theory commitment
M_c	Method commitment
A_c	Aim commitment

THEORY CHANGE—TRIPARTITE OBSERVATION

S_T	Theory of the source
T_T	Theory of transmission
R_T	Theory of the receptor

MATTER

The periodic table of elements is a product of years of investigative activity by chemists working throughout Europe during the eighteenth and nineteenth centuries. During the twentieth century, it becomes a background knowledge (*BK*) resource and a center theory (*C*) used by physicists and chemists to further investigate the composition of atoms and the structure of molecules. The emergence of the knowledge claims that contributed to our view of matter as being fundamentally particulate in nature and as obeying rules of periodicity, leading to the development of the periodic table, is an excellent example for applying our epistemological frameworks.

Characterizing Fundamental Particles

By 1600 mining was an important commercial activity in and among European nations (Laudan, R., 1987). A problem facing the fifteenth- and sixteenth-century mining communities of Europe, however, was the draining of water from the mines. It was a problem because water could not be pumped any higher than 34 feet by a single pump (*IC*). The problem was presented to Galileo Galilei (1564–1642), and he and one of his students, Torricelli (1608–1647), set to work on solving it. Working with a column of water 34 feet high proved to be very difficult. Torricelli had the idea that working with quicksilver (mercury), which is 14 times heavier than water, would be a better method (M_c). When the mercury was poured into a yard-long glass tube that was closed at one end, capped with a finger, and then inverted into a dish of mercury, the mercury in the tube fell, leaving a space between the top of the mercury and the top of the glass column. Torricelli reasoned this space was a vacuum (A_mD), a fact that stood in sharp contrast to Aristotle's notion (*C*) that nature abhors a vacuum.

It was known at the time that air had weight (*IC*), and thus Torricelli reasoned that the column of mercury, which was equivalent by weight to a 34-foot column of water, was held up by the weight of air pushing down on the exposed mercury in the dish (*Fr*). But there was a conceptual problem in claiming that the vacuum itself had special properties that were capable of holding up the mercury. The resolution of this conflict of theoretical commitments (T_c) would take place through carefully designed experiments (M_c) carried out separately by Blaise Pascal (1623–1662) and Otto von Guericke (1602–1686).

Logic would suggest that if the weight of air was indeed the reason the column was supported, then if the weight of air was changed, the height of the mercury should change (N_f). The discovery of this novel fact was made by Pascal, who, as a boy, was fascinated by the news of Torricelli's experiments and performed many of the same experiments himself. In one important experiment he predicted the change in the height of the column of mercury as the glass apparatus was carried up to the top of Puy de Dome, a high mountain in the Auvergne region of France. Although not as dramatic, there was nonetheless a similar change in the height of the mercury when Pascal carried the tube up to the top of a cathedral tower.

The change of theory commitment (T_c) was brought about by the influx of empirical criteria (*E*) from Pascal's experiments and from Otto von Guericke. Von Guericke's methodological commitment (M_c) was the construction of sealed objects, first wooden casks and then bronze hemispheres, to hold vacuums, and the construction of the first air pump to create the vacuum. His evacuated spheres and experiments with the spheres contributed to the empirical criteria (*E*) supporting the theory that air exerted a pressure.

TH_o
Air exerts a pressure that at sea level is capable of supporting a column of water 34 feet high.

BK
1. Air has weight.
2. A vacuum can exist in nature (Torricelli's Vacuum).
3. Water cannot be pumped higher than 34 feet.

IC
1. The height of a column of mercury in a glass tube closed at one end changes with altitude (Pascal's experiment).
2. A sphere evacuated of all air cannot be pulled apart (von Guericke's demonstration).

P
Air exerts a pressure on objects equally and in all directions—there is a sea of air. The upper limit of this pressure is calculated to be 1034g/cm^3 and is called 1 atmosphere. Air confined in a container and compressed is capable of springing back when the confining pressure is removed.

The prediction that air has the property of spring and that it exerts a pressure due to its weight are critically important to the development of core concepts about the composition of matter that culminated in the formation of the periodic table. Our understanding of the composition of matter, as that understanding is reflected in the events that contributed to the development of the periodic table, is considerably a result of experiments with gases.The spring of air and air pressure concepts become background knowledge for the experiments carried out with gases. The theoretical commitment (T_c) is established—vacuums exist and can be made; air exerts a pressure.

Equally important is how these two knowledge claims contributed to the development of methodological commitments (M_c) that made it possible to conduct precise experiments with gases. One is the pneumatic trough method for collecting gases. Today we employ this method by filling graduated cylinders with water, covering the top with a glass plate, and then inverting the cylinder into a trough of water. Given the fact that the cylinder is less than 34 feet tall, the entire column of water is supported. By inserting a tube into the bottom of the cylinder, pure samples of gases can be introduced and collected as the gases displace the water. The device collects gas samples; this we see rather easily. But more important to the growth of knowledge, it is critical to recognize that the gas is being collected at the same pressure—1 atmosphere. This latter observation is made only by invoking our T_c.

The other methodological commitment was the extension of the vacuum pump by attaching it to a glass globe large enough to hold objects for experiments. Clearly, for this to be possible, glass blowing had to have reached a high standard. With this device it was possible to alter the conditions under which experiments took place (that is, either with or without the presence of air—a critical step for determining the conservation of weight in chemical reactions). The development of this apparatus and others by Robert Boyle (1627–1691) not only contributed to a richer understanding of the phenomena associated with the pressure of air, but also helped change the character of science.

Boyle's experiments and his design of apparatus established a new style of doing science, which required detailed descriptions and accurate accounts of events. Boyle, along with other members of the newly founded Royal Society of London (most notably Robert Hooke, who as curator of experiments set up and carried out most of the experiments), made science a public and respectable activity. The aim commitments (A_c) of the gentlemen attending the meetings of the Royal Society were to establish causal significance for the experiments they observed. A large component of this was the ability to repeat and replicate experiments. At this time science was embracing accuracy in measurement, in reporting results, and in describing method, as private knowledge of the laboratory became public knowledge at scientific meetings (Shapin, 1988).

Boyle is best known for his law describing the inverse relationship between the pressure and volume of a gas: pv = constant. Countless numbers of chemistry students who examine the relationship between the pressure and the volume of a gas do so with little appreciation for the commitments (T_c, M_c, and A_c) Boyle and other seventeenth-century scientists held. Devoid of this appreciation for commitments, it becomes difficult to further appreciate the conflict and contributions of Lavoisier (1743–1794) and Priestly (1733–1804). Each of these two men, along with Joseph Black (1728–1799), who isolated CO_2 (fixed air), made significant contributions to the identification and analysis of pure gases, which, in turn, help to move chemistry from a phlogiston view of combustion to an oxygen (breathable air) view of combustion. These men also set the foundation for Dalton's atomic theory of matter and Berzelius' notion of a periodicity of elements—the periodic table of elements.

How we came to know about elements, the combining ratios of elements, the relative weights of elements, and eventually the periodicity of elements was begun and sustained through work and experiments that examined the physical and chemical properties of gases. Only gases could be captured in pure states and held under uniform conditions. When Lavoisier took to weighing gases, however, a problem arose. Were the differences in the weights of gases due to

1. Differences in the number of particles contained in gas samples being collected with equal volumes and at equal pressures; or
2. Differences in the weight of the individual particles that were in the pure gas samples collected with equal volumes and at equal pressures?

The second hypothesis is attributed to Avogadro and is referred to as his EVEN hypothesis—*E*qual *V*olume, *E*qual *N*umber. The equal number is the number of gas particles contained in a sample collected under equal pressure conditions and it is an important piece of background knowledge (*BK*) in the formation of the atomic theory, the discovery of diatomic gases (A_mD), and the calculation of the relative weights of elements (the standard is weight relative to hydrogen, which is a gas).

When first put forth, this claim of there being an equal number of particles was, like all new ideas, a fringe theory (*Fr*). It was a knowledge claim that had been held in question by chemists for several decades. But it was certainly a necessary logical criterion (*L*) for establishing knowledge claims about the relative weights of elements and an atomic theory of matter.

Theory of the Periodicity of Elements

Explaining the development of the periodic table of elements is an important and yet difficult task for science teachers. That such an understanding of matter is based on the investigation of gases is vitally important to the conceptual development of reasoning that a view of matter composed of atoms and molecules is supported by logical, empirical, historical, and sociological criteria. Students can evaluate the status of the knowledge claims put forth by the Atomic Theory of Matter and the periodic table of elements, examined from a developmental perspective, as shown in the following analysis:

Logical Criteria

- The periodic table is a simple unifying idea that places elements in rows and columns according to physical and chemical properties.
- The original table contained vacant slots. The prediction of the element and its properties serves as a test of falsifiability.
- Berzelius originally limited the application of the periodic table to inorganic materials; acid-base-salt interactions are explained by an electrical theory; Wohler's synthesis of urea (A_mD) challenges the idea of vital force explanations for organic substances.

Empirical Criteria

- Test of prediction is satisfied with the discovery of missing elements, for example, germanium.
- Reproducible and quantifiably controlled chemical reactions (for example, acid-base-salt interactions and synthesis of urea) are possible.
- Liquid and solid compounds can be separated into elements (for example, sodium chloride salt into Na and Cl) that obey rules of periodicity.

Sociological Criteria

- Nomenclature and practice of chemistry are unified; diatomic gases are explained; elements are established as fundamental particles.
- New questions are raised about the structure of molecules, the structure of atoms, and rules distinguishing organic chemistry from inorganic chemistry.
- Quantitative analysis (mole calculations) and synthesis of material substances are employed as problem-solving models.
- Elements as fundamental particles with weights relative to the weight of hydrogen make it possible to carry out quantitative analyses.

Historical Criteria

- Alchemy is surpassed as an explanatory model; concepts of intrinsic character of elements (air, water, earth, fire) are eliminated.
- Periodicity of elements accounts for all data gathered by scientists working with gases (i.e., Lavoisier, Priestly, Black, Dalton, Avogadro) and incorporates these results as facts.
- Through the kinetic molecular theory, concepts of elements and molecules having weight are linked with Newton's concepts of motion physics and found to be sound in their explanation of Boyle's law, Charles' law, and Avogadro's EVEN hypothesis.

MOTION

The concepts of motion we attempt to teach our students have a rich developmental history. This history can be exploited to select and sequence concepts toward the goal of students' meaningfully learning motion physics concepts (Nersessian, 1989). Nersessian identifies three stages of development in meaning changes in motion concepts:

1. The medieval theory of motion characterized by local and natural motions
2. Galileo's theory of motion characterized by free fall and circular motion
3. Newton's theory of motion characterized by forces like inertia and gravity

Critical to the examination of the concepts of motion, in any of the three cases above, is the application of the concepts to explaining planetary motion on the one hand, and projectile motion on the other hand. The development of the meaning of motion from Case 1 to Case 3 is one of moving from a multiple perspective of cause to a single explanation for all motion—projectile and planetary—through the Newtonian concept of force.

The Heliocentric View of the Universe

Analyses of the ancients' observations and descriptions of heavenly bodies establish the historical criteria (*H*) that more modern, that is, seventeenth- and eighteenth-century, astronomers had to take into consideration. Focusing on motion, it was clear that the movement of the wanderers—the sun, the moon, Mercury, Venus, Mars, Jupiter, Saturn—was very different from the movement of the stars. The aim commitment (A_c) was to predict as accurately as possible the position of the planets. The method commitment (M_c) was to geometry and numerical harmony. The theoretical commitment (T_c) was to a geocentric view of a bounded perfect universe composed of crystalline spheres.

The legacy of the alteration of the geocentric view of the universe to the heliocentric view can be understood in terms of changes in A_c and M_c, bringing about changes in T_c. What is intriguing is how much evidence, in the form of observational data of planet positions, calculations of the distance to planets, and calculations of the motion of planets, was needed to bring about such changes.

The A_c to predict and explain the position of planets was challenged and extended by

1. The retrograde motion of Mars (A_mD)
2. Measurements of the relative size of the moon, earth, and sun
3. Calculations of the shape and size of the earth and the distance from the earth to the moon and the sun

Establishing the size of the solar system (*E*) was critical to understanding the conceptual shift from a geocentric to a heliocentric model of planetary motion. Any suggestion, however, that the earth, along with the other planets, turned or traveled around the sun had to account for the lack of parallax shift (A_mD) among the stars and the lack of visible effects of motion on the earth (such as wind in the face). Each of the measured patterns of motion—seasons, precession of equinoxes, retrograde motion of planets, solar and lunar eclipses, phases of the moon—became *IC* contributing to a new view of the solar system. Such measurements were nonetheless unable to alter the T_c to a heliocentric model of the universe; even as late as Tycho Brahe (1546–1601) astronomers were committed to a geocentric model of the universe. In fact, Copernicus originally set out to modify Ptolemy's system of the equant to preserve Aristotle's notion of uniform motion.

The approach to understanding the change of planetary models is through the changes brought about to the *BK* of the geocentric model. The significance of the retrograde motion of Mars is that it challenged the *BK* of circular motion. The significance of Aristarchus' calculation of the distance to the sun is that it began to alter the *BK* of a finite universe. Copernicus' calculation of the distance to the planets, employing the siderial and synodic periods of the planets, further altered the idea of a finite universe.

TH_o
The earth is at the center of the universe and all celestial objects move around it.

BK
1. The stars and planets are contained in a set of crystalline spheres set in motion around the earth (thus the size of the universe is finite).
2. All motion is circular.
3. All motion is uniform—at a constant velocity.

IC
1. Mars displays retrograde motion.
2. The stars never display parallax.
3. Accurate patterns of motion exist for the sun, moon, and planets.

P
The position of the planets, sun, stars, and moon can be predicted accurately, thus allowing for forecast applications to agriculture, religious ceremonies, and navigation.

In understanding the change from the geocentric to the heliocentric model, the critical steps are those taken to dismantle each of the *BK* and *IC* claims for the geocentric theory. As mentioned above, a critical element in understanding the theory change is the steps taken to measure the size of the universe. What occurs is intriguing. Alterations to and improvements in the manner in which observations of celestial positions were made and recorded provided anomalous data (A_mD) to attack the *BK* and *IC* above. The A_mD evidence became counterexamples to the claims of the geocentric theory testing argument. Thus, Galileo's (1564–1642) telescopic observations of the phases of the moons of Jupiter challenged (A_mD) claim *BK 1*. Copernicus' calculation of the size of orbits and speed of planets in the orbits contributed to the dismantling (A_mD) of *BK 1* and *IC 2*, because his calculations began to call into question the finite model of the universe.

Copernicus' move of placing the sun at the center position sought to modify Ptolemy's system of the equant so as to preserve *BK 3*—uniform motion. In so doing, Copernicus introduced a new model (*Fr*)—the heliocentric model. The significance of the new model is it accounted (N_f) for the retrograde motion of Mars without appealing to major and minor epicycles. A thousand-year-old empirical problem had been resolved! It is also a consistent model for explaining the motion of the sun and stars (*L, E*).

Further evidence for the heliocentric model was unintentionally provided by Tycho Brahe through his very accurate and daily recordings of planetary positions. Brahe supported the geocentric model of the universe, but his measurements, which, it must be stressed (M_c), were accurate to within 2 minutes of arc (1° of arc = 60 minutes), contributed significantly to the amount of A_mD that is unexplained by the geocentric model.

Kepler (1571–1630), an assistant to Tycho Brahe, demonstrated the importance of M_c and A_c to the growth of scientific knowledge. Kepler held that because the observations of the heavens were accurate to within 2 minutes of arc, any model for predicting the position and the motion of the heavens should be as accurate (M_c). The A_c for Kepler, in common with Brahe and Copernicus, was accuracy in predicting the position of planets and describing the motion of the planets—clearly not unrelated goals.

The *P* of the theory testing argument for each model remains the same—accurate predictions. Unlike Brahe, however, who was committed to a geocentric model, Kepler was sympathetic to Copernicus' heliocentric model and thus was willing to toy with established knowledge claims to improve the accuracy to the within-2-minutes-of-arc goal. Through Kepler's careful analysis of Brahe's data—*BK 2* (circular motion), *BK 3* (uniform motion), and *IC 2* (no star parallax) are undone. Kepler's laws of planetary motion provide a new set of *IC* for the heliocentric TH_o, and each of the laws, in turn, describes the character of the movement. Thus

TH_o
The sun is the object around which the planets move.

BK
1. The stars are a very large distance away and remain stationary.

2. All motion is elliptical with the sun at a focus.
3. All motion is nonuniform—velocity changes; motion is governed by equal areas over equal units of time.

IC

1. Mars displays retrograde motion.
2. The stars never display parallax.
3. Accurate patterns of motion exist for the sun, moon, and planets.
4. The period of any planet's orbit, squared, divided by the distance from the sun, cubed, is equal to a constant.
5. The moons of Jupiter orbit Jupiter.
6. Phases of Venus exist.

P

The position of the planets, sun, stars, and moon can be predicted accurately, thus allowing for forecast applications to agriculture, religious ceremonies, and navigation.

Mechanics of Motion: From Natural Place to Force

The heliocentric model of the solar system, while certainly enhancing our ability to predict and understand what is taking place, does little to inform us about how and why it works this way. The appeal of circular motion and uniform motion is a derivative of Aristotle's theory of motion. For Aristotle motion was either natural (up or down) or forced (horizontal); this was the dominant view (*C*). Following the revival of the study of science after the Dark Ages, a revival attributed to Thomas Aquinas (1225–1274) and stimulated by Christian theology, Aristotle's ideas about projectile motion were found to have many contradictions, such as Zeno's paradox. Questions about forced or projectile motion were addressed by many, but Galileo is credited with changing our basic theoretical models of motion. A critical element (*BK*) in the development of a universal principle of gravitation is the separation of motion and forces into component elements (M_c). The analysis of compound projectile motion into two straight-line components was initiated by Galileo. From this analysis we acquire another critical element (*BK*) in the development of the concept of gravity, namely, acceleration.

The bridge from Galileo's concepts of motion to Newton's laws of motion and principle of gravity begins with concrete, measurable concepts of motion and finishes with abstract nonintuitive concepts of motion, mass, and force. If the foundation of concrete concepts is weak or faulty, then the abstract ideas of centripetal force, of mass being different from weight, and of force implying acceleration, and vice versa, will not meaningfully link to prior knowledge claims. The explanation of why planets orbit the sun and the moon orbits the earth is grounded in an analysis of projectile motion carried out through strict experimentation.

Through a series of thought experiments (*L*) and data collection experiments (*E*) employing pendulums and inclined planes (M_c), Galileo was able to show that projectile motion along a vertical plane was fundamentally different from motion along a horizontal plane. The characterization of vertical motion as uniform accelerated motion was established by experiment (*E*). Using an inclined plane to slow down the action of vertical fall, Galileo was able to substantiate his reasoned claim (*Fr*) that a falling object would speed up as it fell. By rolling balls down a precisely crafted length of hardwood, he was able to establish that the total distance traveled is equal to the square of the time (*E, IC*), or $s = \frac{1}{2}at^2$; where *s* is the distance, *a* is the acceleration, and *t* is time. Stated verbally, over equal units of time, there is a uniform increase in the distance traveled. Hence, in each time unit in the table below, the object travels two more units of distance than in the time before.

Time	*Distance/Total Distance*	*Change in Distance*
		per time unit
1	1/1	
2	3/4	2(3 − 1)
3	5/9	2(5 − 3)
4	7/16	2 (7 − 5)
5	9/25	2 (9 − 7)

The important point here is that vertical motion is a different kind of motion from horizontal motion. Empirical criteria (*E*) established the characterization of vertical motion as uniform accelerated motion: *equal time, equal increase in distance traveled.* Logical criteria (*L*), however, established the characterization of horizontal motion as uniform motion: *equal time, equal distance traveled.* Using the inclined plane model and the motion of a pendulum, Galileo reasoned that if motion down the plane was uni-

FIGURE 7.2 Depiction of Galileo's Pendulum Experiment[a]

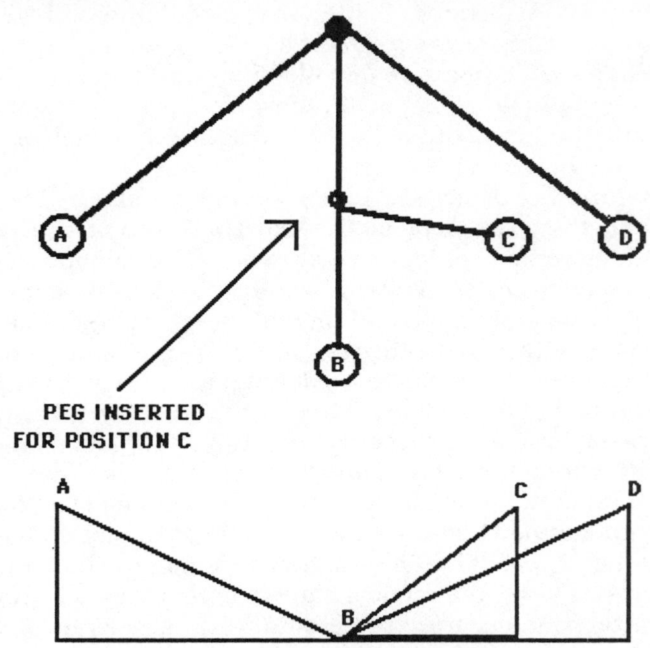

[a] In this experiment both positions C and D will achieve the same height A. Positions C and D have the same height as A but traverse different horizontal distances to point B. This description is clear from examining the base of the triangles in the lower figure.

form accelerated motion, then motion up the plane would be uniform decelerated motion, and motion on a horizontal plane would be neither accelerated nor decelerated.

Galileo was next able to demonstrate through experiments with a pendulum that the two motions acted independently of each other. That is, changes to the quality of horizontal motion had no effect on vertical motion (herein lie the roots of vector constructions). Thus, in an experiment that kept the vertical height of a pendulum constant but changed the horizontal distance traveled on the right-hand swing (see Figure 7.2), the resultant left-hand swing was the same, that is, achieved the same height (*IC*). Galileo frequently substituted the action of balls on inclined planes for the action of a pendulum in his experiments. This informs us as teachers about the need to use both these devices in our demonstrations and laboratory investigations. Galileo reasoned that a ball rolled down an inclined plane at an angle of 45° will, ignoring friction, achieve the same velocity as a ball rolled down an inclined plane of the same height at an angle of 10°. Inasmuch as the vertical action is mutually exclusive of the horizontal action, the angle is unimportant.

The separation of projectile motion into horizontal and vertical components that are independent of each other is a critical step (*BK*) in the development of ideas about why and how motion occurs. We can reconstruct the representation of Galileo's thought experiments by calling on the pendulum and inclined plane to demonstrate his points. Clearly, Galileo's description of motion as both vertical and horizontal was needed to develop the idea of gravity.

The separation of motion and the separate qualities of the motion set the stage for the work of Huygens (1629–1695) on momentum and of Newton (1642–1727) on the laws of motion and concepts of force. In fact, the separation of the mass of an object from its weight is understood employing the separation of motion along the horizontal and vertical plane—mass acts along the horizontal plane; weight acts along the vertical plane. Another anchoring concept for Newton is an accurate representation of acceleration, because $F = ma$ states that the presence of an acceleration implies a force and vice versa. This simply stated relationship between force and acceleration is the cornerstone or anchoring concept for the theoretical hypothesis (TH_o) that planetary motion is governed by an inward centripetal force called gravity.

The concept of motion being governed by gravity is one students accept but are frequently unable to explain. The path of explanation, history informs us, involves both planetary motion and projectile motion. Explaining the dismantling of Aristotle's concepts of circular and uniform motion by Kepler and Galileo is a critical task for teachers in presenting the rational development of Newton's theory of motion and theory of gravity.

NOTES

[1] The first three storylines described in this and the next chapter are based on the Foundations of Science course offered at Hunter College of the City University of New York and developed by Alfred Bennick, David Lavallee, Ezra Shahn, Fred Szalay, and Richard Duschl.

Chapter 8

Applying the Growth of Knowledge Frameworks —Life Science and Earth Science

The foundations of science have their deepest theoretical roots in the physical sciences. Members of ancient civilizations—for example, Ptolemy, the Babylonian astronomers, Anaximander—began to offer explanations based on observational evidence. This tradition of obtaining observational evidence has, as we have already learned, developed in time to include increasingly more sophisticated techniques and technologies for carrying out scientific investigations. In Chapter 5 the idea was introduced that any observation has three components—source, transmission, and receptors—each with a set of theories that establishes standards for judging "what counts." The traditions of science and analyses of investigative methods of science have, until very recently, focused principally on developments in the physical sciences. For a variety of reasons that are too elaborate to engage in this text, our earliest theories of science emerged first from physics/astronomy and then from chemistry. This is not to say, of course, that individuals did not investigate matters of geology and biology. They did. Rather, the point is that the fields of biology and geology were not directed by theories of their own disciplines until the twentieth century.

Kuhn (1962/1970) characterizes the difference between theoretically driven science and atheoretically driven science as mature and immature science, respectively. His claim is that in each scientific discipline there comes a time when its practitioners reach consensus on what are to be the central theories that guide the discipline. Both physics and chemistry arrived at central theories hundreds of years earlier than biology and geology. For biology, the transition from immature science to mature science took place in the 1920s to 1930s with the great synthesis of mechanisms of natural selection with mechanisms of genetic selection. For geology, the transition took place in the 1950s and 1960s with the introduction of evidence of plate tectonics. In this chapter, we turn our attention to the development of knowledge claims in the life sciences and earth sciences.

ACQUIRING A HISTORY OF LIFE ON EARTH

The development of the view that the earth and life on earth have an elaborate history that far exceeds the presence of humans owes a great deal to geology and theories about the age of the earth. What evidence and reasoning allow us to date the age of the earth as hundreds of thousands of years old and thus enable us to invoke a model of life on earth governed by a theory of evolution? The use of the word *enable* in the previous sentence is important for it suggests correctly how the development of scientific theories is dependent on the knowledge claims (*BK, S;* refer to Figure 7.1 for codes) of other scientific fields of inquiry.

The development of the modern theory of biological evolution is an excellent example of the interdisciplinary frameworks that contribute to the growth of knowledge in science. To ignore these interfield relationships is to ignore the sociological criteria (*S*) so important in the evaluation of scientific theories. Before Charles Darwin (1809–1882) was able to lay down the beginnings of a theory of evolution in the form of his principle of natural selection, it was necessary for him to be committed to views of plutonism, uniformitarianism, extensive diversity of life, and an earth that was hundreds of thousands of years old.

The growth of knowledge in modern biology owes a great deal to the work of geologists like James Hutton, Charles Lyell, John Playfair, John Hall, and William Smith. These founders of geology (Geikie, 1905/1962; Laudan, R., 1987) made it possible to burst the bubble of time about the age of the earth and thereby allows others to conceive of natural processes that took previously unimaginable amounts of time and generations to occur. The growth of modern biological knowledge also owes a debt to naturalists and taxonomists like Carl Linnaeus, Georges Cuvier, Georges Louis Leclerc Buffon, and Jean Baptiste Lamarck. A critical step in the development of ideas about evolution is the shift in thinking from biological individuals to biological populations; that is, biological change occurs in populations, not in individuals. Without the extensive catalogues (*E, IC*) of life on earth and the diversity of life on earth that were amassed in European museums, evolution as a mechanism that takes place within populations would be difficult to conceive.

Because of the importance of a very old earth for a theory of evolution, a strong argument can be made that England was the only location where the intellectual climate was capable of supporting an explanation like Darwin's theory of evolution. The chief warrant for making this knowledge claim is that England and Scotland were the birthplace of modern geology. James Hutton (1726–1797) introduced two ideas (*Fr*) that challenged the established views (*C*) about the formation of rocks. Hutton proposed that rocks were not the exclusive result of deposition by or in water, but rather that some rock formations owed their existence to intrusive mechanism involving molten rock. The laying down of rock by water was the preferred theory of rock formation among German mineralogists, particularly Abraham Gottlob Werner (1750–1817). Werner advocated that all rocks were the result of action by or in water. This view was called neptunism and stood in sharp contrast to Hutton's plutonism.

Hutton is a key player in the growth of modern biological thought. R. Laudan (1987) suggests, for example, that Hutton brought very new perspectives to geological processes. First, Hutton's theology—deism—allowed him to think in terms of a cyclic, causal theory of the earth. Second, his exposure to Scottish interpretations of heat theory had him thinking differently than mineralogists from Germany. The theology of deism and conceptions of heat are important (*BK*) to Hutton's principle of uniformitarianism.

> As a deist, Hutton asserted the existence of a deity, but denied the evidences of revealed religion. In particular, like other deists, he dismissed the Christian story, laid out in the Old and New Testaments, as unworthy of serious consideration. . . . Christian geologists thought that the Bible, like other ancient writings, offered invaluable direct testimony about the earth's past. Deists had no use for such endeavors. Biblical stories were myths, not to be taken seriously by the natural philosopher. . . . Theological differences between Christians and deists determined not only the evidence they considered relevant to theories of the earth but the very *kind* of theory that they advanced. (R. Laudan, 1987, p. 115)

Hutton, therefore, was free to consider alternative scenarios to the history of the earth, and he chose to focus on a view that allowed for recurring cycles of decay and renovation. The steady-state destructive processes of erosion provided the materials for the renovating constructive processes of deposition and plutonic intrusion. The resolution of the theoretical debates between neptunism and plutonism and between catastrophism and uniformitarianism are crucially important steps in the rational development of knowledge claims made and synthesized by Charles Lyell (1797–1875) about geology and the age of the earth.

TH_o
The processes of rock formation are cyclic in nature, involving the destructive action of water and gravity (erosion) and the constructive action of water and heat (deposition).

BK
1. Deist theology allows for natural processes to govern the mechanisms of earth formation. Biblical accounts of creation are dismissed.
2. The powers of gravitational matter (Newton) and of heat (Boerhaave and Black) provide mechanisms to drive the processes of earth decay and renovation.

IC
The British regional geology provided numerous examples of decay and renovation.
1. Unconformities (tilted rock overlain by horizontal strata).
2. Granite intrusions.
3. Folded and faulted rock formations.

P
A theory of the earth is best explained by principles of plutonism and uniformitarianism.

Lyell drew on the principles of plutonism and uniformitarianism, and the accompanying background knowledge cited above, in a study of the geology of Mt. Etna in Italy. It was here that the empirical criteria (*E*) were in place for Lyell to propose that Mt. Etna was the result of the renovating action of heat that Hutton had proposed. Using lava flows, fossil formations, and Hutton's principle of uniformitarianism, it was also apparent to Lyell that Mt. Etna had taken enormous amounts of time to reach the height and breadth it had achieved. Lyell succeeded in establishing with his trip to Mt. Etna that the earth was much, much older than anyone had conceived of before. Lyell incorporated these ideas of the enormous age of the earth and of uniformitarianism into his *Principles of Geology*, a two-volume set of geological knowledge that Charles Darwin read and endorsed.

Also important to Darwin's development of natural selection as a mechanism for evolution was the work carried out in classification and the development of taxonomies. The catalogues of life forms established by Carl Linnaeus (1710–1778)—the binomial system of naming plants and animals—contributed important evidence to establish the concepts of diversity and species. Evidence that life on earth is varied and diverse led many to entertain ideas of transmutation and other ideas of how life form came to be similar to or different from other forms. An aim commitment (A_c) that emerged in the biological sciences was how to explain the diversity of life in nature, which also reflected an increasing complexity in life forms. The study of stratigraphy and fossil sequences, especially by Baron Georges Cuvier (1769–1832), for example, not only helped advance Huttonian ideas of uniformitarianism, but also gave the diversity and complexity of life a historical dimension as well. Specific empirical evidence (*E*) for the diversity of life included:

Collections of living specimens
Collections of fossils
Descriptions of the biogeographic distribution of life forms
Studies of the comparative anatomies of life forms
Studies of the embryonic development of life forms

This evidence led some early evolutionists, for example, Lamarck (1744–1829), to propose theories of change that operated via mechanisms within the individual organism. The task of comparing Lamarck's explanation for biological diversity with Darwin's explanation is simplified if we highlight the differences in the background knowledge (*BK*) and initial conditions (*IC*) between the two explanations.

With respect to *BK*, for Lamarck the agent of change was a mechanism within the individual, whereas Darwin focused on a mechanism affecting populations. Next, for Lamarck the struggle for survival was one of an individual adapting to the environment in terms of acquiring food. Darwin's concept of struggle for existence included survival, but in the special sense that it allows an organism to reproduce. Thus it is via selection of individual organisms to reproduce that Darwin employed the terms *survival* and *adaptation* of a population. Lamarck, on the other hand, talked of survival and adaptation of individuals occurring without a role for reproduction. One important consequence, then, in the rational development of a theory of biological evolution is that the meaning of the term *species* changed. Prior to Darwinian natural selection, species were defined as organisms with similar traits; that is, using Linnaeus' binomial classification schema. What Darwin's theory of natural selection provided, however, was a new method commitment (M_c), namely, defining species in terms of populations. Thus, species were no longer defined by common characteristics, but by the physical and behavioral capacity to reproduce in nature.

The M_c to populations was very significant. It emerged very prominently in Mendel's experimental design with peas, which laid the foundation for modern genetics. It was sustained in subsequent applications of mathematical theory to the study of population genetics so important in the development of twentieth-century agricultural practice. There is another distinction between Lamarck, a Frenchman, and Darwin, an Englishman, which we have already alluded to—the commitment Darwin had to British geology. In the argument for a theory of natural selection, the initial conditions (*IC*) provided by geology about the age of the earth made it possible to conceive of an explanation for the evolution of life based on a mechanism involving selective reproduction over successive generations within a population.

We arrive then in the twentieth century and find scientists using the ideas of Darwin, Mendel, and the more recent population genetic thinkers like G. H. Hardy and W. Weinberg as different theory commitments (T_c) for guiding observation and for determining what counts as evidence (S_T, T_T, R_T). Darwinian naturalists advocated field research to better understand the dynamics of natural selection. Experimentalists supporting Mendel advocated laboratory studies to better understand the dynamics of genetics. This period of the growth of knowledge in biological sciences is a classic one in which dissensus existed between those scientists who supported Darwin's natural selection as the mechanism for evolution and those who supported Mendelian genetics as the mechanism. Here we find two fringe theories (*Fr*) competing with each other.

It was the mathematical population genetic thinkers and their use of statistics that brought about a consensus of theory commitment (T_c) in the biological sciences. The source observational component (S_T) and the receptor observational component (R_T) for individuals working in population genetics, such as R. A. Fisher and Sewell Wright, were guided by mathematical theory. For example, Wright "devised a mathematical technique of path co-efficients to estimate the degree to which a given effect was determined by each of a number of causes (1917–1921). He became convinced that the interaction of genes was an important factor in evolution" (Brush, 1988, pp. 153–154). Here is a case where theories of the source and theories of the receptor led to theories of the transmission of evidence, which, of course, is genes. New interactions of genes, for example, cross links, were proposed to explain the predicted results of population theory.

Population theory changed what was considered evidence among biologists and thereby changed the practice of research in the field. The mathematical theory established a quantitative framework for the interpretation of data, identifying relevant parameters, suggesting experimental designs, and generating testable hypotheses. How else does one make sense of the decades-long research, begun in Russia under the guidance of S. S. Chetverikoff (1880–1959) and carried out at Columbia University by Theodore Dobzhansky (1900–1975) on fruit flies. This research helped settle the debate between external forces of evolution (reproductive isolation) and internal forces of evolution (genetics) that ultimately led to what has come to be known as the great synthesis in modern biology (Mayr, 1982). The great synthesis established a more comprehensive explanation (Ft) for biological diversity and complexity by joining genetic mechanisms for changes with natural mechanisms.

The development of the modern understanding of the theory of evolution—neo-Darwinian evolutionary theory—is a two-century-long investigation into the age of the earth and the diversity and complexity of life on earth. The growth of knowledge associated with the emergence of a theory of evolution can be simplified to two critical episodes of unification or synthesis. Darwin was the first synthesizer of knowledge from geology and biology. New evidence (A_mD) introduced by Mendel and the mathematical population geneticists could not be ignored and forged a dissensus among biological scientists about the proper role of external and internal mechanisms of evolutionary forces. The next synthesis was led by Dobzhansky and paved the way for our contemporary foray into molecular biology. Darwin's unification and Dobzhansky's unification occurred approximately 100 years apart. Each embraces a legacy of scientific activity that only a developmental perspective of theory change can capture. As science teachers, we come to see how important geology, population thinking, reproductive isolation as a criterion for defining species, and mathematical theory are to the growth of knowledge in biology.

MOVING PLATES

The development of the theory of plate tectonics is a twentieth-century phenomenon. An analysis of the data sources that contributed to drafting the view that the earth is composed of plates moving relative to one another reveals that here, too, as with the theory of evolution, the growth of knowledge was an interfield event.

The historical story of the theory of plate tectonics owes more initially to the contributions of physicists studying magnetism and oceanographers studying the bottom of the ocean than to geologists mapping rock structures and rock formations. The period of unification was the 1950s and 1960s, the place, once again, was primarily England, with substantial input from Japanese researchers and U.S. oceanographers. The critical events involve classical experiments in physics, advances in technology, and political events.

Magnets Point the Way

By the beginning of the twentieth century, the existence of the earth's magnetic field had been known and studied for over two thousand years. William Gilbert (1544–1603) published his important work on magnetism in 1600. Shortly thereafter, ships and magnetic observatories were outfitted with compasses capable of measuring both the declination (left–right) and inclination (up–down) displacement of a needle. Magnetic compasses to find north had been in use for over a thousand years before 1600, but with the more sensitive and accurate instruments it was possible to determine both longitude and latitude so precisely that, after several hundred years of data collection, variations in readings at single sites were being discovered.

Although much was known about the effect of the earth's magnetism, little was known about the cause. In the middle of the nineteenth century Maxwell had demonstrated the relationship between electric and magnetic forces—one capable of producing the other—in his development of the electromagnetic theory. But this excluded the natural magnetism of the earth. About 100 years later, a team of British scientists headed by P.M.S. Blackett decided to make a serious assault on the problem of finding a cause for the earth's magnetic field. The prevailing fringe theory (*Fr*) to explain the earth's magnetic field in 1950 was the dynamo theory. At this time it was known that the core of the earth was surrounded by a liquid (*BK*). It was also known that the earth rotated and that the core of the earth was much more dense than the surface. Blackett and his team of researchers hypothesized that the rotation of a dense object would produce a magnetic field.

To put this theory to a test, it was necessary to construct an instrument that could measure very weak magnetic fields. The instrument constructed was the astatic magnetometer. The experiments were run by rotating cylinders of gold, and the results were negative—rotating objects did not produce magnetic fields. However, the astatic magnetometer proved to be a valuable instrument for measuring the remanent magnetism of iron-bearing rock samples. What emerged was a new and powerful observational technique in which theories of the source (S_T) based on concepts from magnetism and theories of the receptor (R_T) also based on a theory of magnetism were applied to a new domain—rock samples. For a detailed presentation of this intriguing episode in the history of science, readers are referred to the book *Debate About the Earth* by Takeuchi, Uyeda, and Kanamori (1970). What follows here is a brief outline of the salient features and events that contributed to the evidence that helped revive Wegener's theory of continental drift and also gave rise to the theory of plate tectonics.

The Theory of Plate Tectonics

The science of remanent magnetism in the 1950s was advanced by Japanese researchers at the University of Tokyo and also in France. Scientists had been analyzing the mineral composition of magma flows for several decades and discovered that the iron minerals in the magma flows would act like individual magnetic needles and freeze along the prevailing magnetic flux lines characteristic of that location on the earth. They also found that lava flows from the same location did not always have the same magnetic directions. The rocks on the earth were a repository of the earth's magnetic history, freezing iron minerals with unique declination and inclination orientations. This phenomenon is called fossil magnetism, and the study of it, paleomagnetism.

The British scientists were familiar with the paleomagnetic research being conducted by the Japanese. Perhaps this fossil magnetism data would shed light on a theory of the earth's magnetism, so they began measuring the remanent magnetism of rocks of different ages found in England. What is significant here is that prior to the development of the astatic magnetometer, only igneous rocks could be examined for remanent magnetism. With the astatic magnetometer it was possible to pick up the remanent magnetism in all types of rocks—igneous, sedimentary, and metamorphic. Anomalous data (A_mD) about the magnetic history of the earth was mounting rapidly. Significantly, this data was not accounted for by any existing theory (*H*). The empirical criteria (*E*) that were amassed are worth examining, for they led physicists to test and revive the theory of continental drift.

The geologic record of England spans the paleozoic, mesozoic, and cenozoic eras. The study of these rocks revealed that the magnetic north orientation was different over geologic time (A_mD). The idea that the magnetic north pole of the earth could shift had been entertained before as a possible explanation for the secular variation in magnetic readings recorded at magnetic observatories. It was proposed as the theory of polar wandering (*Fr*).

The competing fringe theory to explain the changing rock magnetism was Alfred Wegener's continental drift explanation. The logical and empirical criteria for the drift of continents (*L, E*) were based on the distribution of continental land masses and on rock and fossil types. What was missing was the viable background knowledge about the interior of the earth that would support an explanation like moving continents. Recall the pattern first presented in Chapter 5 (p. 72).

Not until a change in *BK* came about could the *P* of drifting continents be advanced. The sociological criteria (*S*) for evaluating theories in 1955 were very different from those in 1926. The study of the behavior of seismic waves suggested that the earth was not entirely solid. To explain the patterns of data being collected by seismographs, a partially liquid-core model of the earth had to be adopted. Once the veil of an entirely solid-core earth had been lifted, seismologists also found it prudent to propose that the upper section of the mantle was more accurately described as a plastic substance than as a solid.

Shifting the *BK* from a solid-core earth to a partially liquid-core earth was the critical turning point that enabled drift theory to be considered a viable explanation for the varying geomagnetic patterns found in rocks on the British Isles. The two competing explanations (*Fr*) were polar wandering and continental drift. To determine which was a better explanation, two tests of rock remanent magnetism were made; one to test the polar wandering thesis was carried out in North America and the other, to test Wegener's drift thesis, was conducted in India.

If Wegener's idea of continental drift was correct, then a particular pattern of rock magnetism would be found in India (N_f). Wegener had proposed that all the continents were once joined together in the southern hemisphere and subsequently moved to their present positions. Among these land masses, India had traveled a path almost due northward. If drifting of India had taken place, then the magnetic inclination recorded in iron minerals in successively younger rocks would reflect the change in latitude. Such a pattern was found (*E*), supporting the idea of continental drift.

The test of rocks in North America was set up to examine the theory of polar wandering. If the theory of polar wandering was correct, then all rocks formed in the northern hemisphere at the same time should point to the same magnetic pole position (N_f). Rocks from each of the cenozoic time periods (Cambrian, Ordivician, Devonian, Triassic) tested in England were sampled in North America, and a different polar wandering path was found. Additionally, when the magnetic pole position was held constant and the land masses were allowed to drift, the fit supported Wegener's thesis (*E*).

The studies of continental rocks in England, India, and North America produced a set of data that indicated that remanent magnetic directions varied with age. Even more surprising, however, was the discovery in the paleomagnetic record of periodic reversals of the earth's magnetic field, which took place about every 500,000 years (A_mD). These reversals became critically important empirical/theory-of-source criteria (E, S_T) for the advancement of the theory of plate tectonics. Taking the fact of reversals (*E*) and combining it with the fringe theory (*Fr*) of sea-floor spreading, it was possible to explain the symmetric rock patterns found on the opposite flanks of the mid-ocean ridges. Thus, the theory of plate tectonics emerged from the work of physicists studying magnetic theories of the earth.

Here is a case in which theory change was initiated by shifts in methods (M_c) and aims (A_c) before shifts in theory commitment (T_c). Employing a context-of-development approach to the study of plate tectonics requires that attention be given to the important role British physicists played in the revival of the idea of continental drift. These scientists found that the astatic magnetometer could be used to document the paleomagnetism of rocks. Their subsequent studies of continental rocks in England, India, and North America produced a set of anomalous data that challenged prevailing thinking about geologic dynamics. Employing a context-of-development point of view, we find that a change in method (using the astatic magnetometer to measure rocks) led to a change in cognitive aim (physicists turning their resources and efforts away from a classical problem of magnetism to the problem of continental drift) that, in turn, led to a total restructuring of our knowledge of the earth.

Employing an instructional decision-making model that seeks to select and sequence concepts that explain how we have come to know about plate tectonics would place equal emphasis on methods, aims, and theory commitments. Whereas final form presentation of science would emphasize the last revision of the restructured knowledge (sea-floor spreading to plate tectonics employing magnetic reversal patterns on the ocean floor), the developmental presentation focuses attention on the *BK, IC, L, E, H,* and *S* that caused a shift in S_T, T_T, and R_T. In this way, the rational evolution of the theory of plate tectonics emerges.

REFERENCES

INDEX

ABOUT THE AUTHOR

References

Albritton, C. C. (1963). *The fabric of geology*. Reading, MA: Addison-Wesley.

Alvarez, L. W., Alvarez, W., Asaro, F., & Michel, H. (1980). Extraterrestrial causes for the cretaceous-tertiary extinction. *Science, 208,* 1095–1107.

Anderson, C., & Smith, E. (1986). Teaching science. In V. Koehler (Ed.), *Education researchers handbook.* New York: Longman.

Benson, G. (1989). The misrepresentation of science by philosophers and teachers of science. *Synthese, 80:*1, 107–119.

Bransford, J. D., & Stein, B. S. (1984). *The ideal problem solver: A guide for improving thinking, learning, and creativity.* New York: W. H. Freeman & Co.

Bronowski, J. (1974). *The ascent of man.* Boston: Little, Brown.

Bruner, J. (1960). *The process of education.* Cambridge, MA: Harvard University Press.

Brush, S. (1988). *The history of modern science: A guide to the second scientific revolution, 1880–1950.* Ames: Iowa State University Press.

———. (1976). *The kind of motion we call heat.* New York: North Holland/American Elsevier.

Bybee, R. (1987). Science education and the Science-Technology-Society (S-T-S) theme. *Science Education, 71:*5, 667–683.

———. (1984). Global problems and science education policy. In R. Bybee, J. Carlson, & A. McCormack (Eds.), *Redesigning science and technology education—1984 NSTA Yearbook.* Washington, DC: National Science Teachers Association.

Carey, S. (1986). Cognitive science and science education. *American Psychologist, 41,* 1123–1130.

———. (1985a). *Conceptual change in childhood.* Cambridge, MA: Bradford Books/M.I.T. Press.

———. (1985b). Are children fundamentally different kinds of thinkers and learners than adults? In S. Chipman, et al. (Eds.), *Thinking and learning skills,* Vol. 2, 485–517. Hillsdale, NJ: Lawrence Erlbaum Associates.

Carter, W. E. et al. (1984). Variation in the rotation of the earth. *Science, 224,* 957–961.

Champagne, A. (1988, April). A psychological model for science education. Paper presented at the meeting of the American Educational Research Association, New Orleans.

Cohen, I. B. (1985). *Birth of a new physics*. New York: Norton.

Conant, J. (1957). *Harvard case histories in experimental science*, Vols. 1 & 2. Cambridge, MA: Harvard University Press.

Connelley, F. M. (1972). The function of curriculum development. *Interchange, 3*, 164.

Connelley, F. M., & Finegold, M. (1977). *Patterns of enquiry project—Scientific enquiry and the teaching of science*. Toronto, Ontario: Institute for Studies in Education Press.

Crane, L. T. (1976). *The National Science Foundation and precollege science education: 1950–1975*. Washington, DC: U.S. Government Printing Office. 61–660.

de Queiroz, K. (1988). Systematics and the Darwinian revolution. *Philosophy of Science, 55*:2, 238–259.

DeRose, J., Lockard, J. D., & Paldy, L. P. (1979). The teacher is the key: A report on three NSF studies. *The Science Teacher, 46*:4.

DeWitt, N. (1955). *Soviet professional manpower—Its education, training, and supply*. Washington, DC: National Science Foundation, U.S. Government Printing Office.

Durbin, P., Ed., (1980). *A guide to the culture of science, technology, and medicine*. New York: Free Press.

Duschl, R. (1988). Abandoning the scientistic legacy of science education. *Science Education, 72*:1, 51–62.

———. (1987). Causes of earthquakes: An inquiry into the plausibility of competing explanations. *Science Activities, 34*, 8–14.

———. (1985). Science education and philosophy of science: Twenty-five years of mutually exclusive development. *School Science and Mathematics, 85*, 541–555.

Duschl, R., & Wright, E. (1989). A case study of high school teachers' decision making models for planning and teaching science. *Journal of Research in Science Teaching, 26*, 467–501.

Dutch, S. I. (1982). Notes on the fringe of science. *Journal of Geological Education, 30*, 6–13.

Easley, J. (1959). The Physical Science Study Committee and educational theory. *Harvard Educational Review, 29*:1, 4–11.

Eisner, E. (1985). *The educational imagination: On the design and evaluation of school programs*, 2nd ed. New York: Macmillan.

Eldredge, N. (1981, April 4). Creationism isn't science. *The New Republic*, pp. 15–17, 20.

Frankel, H. (1983). The development, reception, and acceptance of the Vine-Matthews-Morley hypothesis. *Historical Studies in Physical Science, 13*, 1–39.

Geikie, A. (1905/1962). *The founders of geology*. Reprinted. New York: Dover.

Giere, R. (1988). *Explaining science: A cognitive approach*. Chicago: University of Chicago Press.

———. (1984). *Understanding scientific reasoning*. 2nd ed. New York: Holt, Rinehart and Winston.

Gillispie, C. C. (1960). *The edge of objectivity*. Princeton, NJ: Princeton University Press.

Goodlad, J. (1966). *The changing school curriculum*. New York: The Fund for the Advancement of Education.

Grandy, R. (Ed.). (1973). *Theories and observation in science*. Atascadero, CA: Ridgeview.

Gunstone, R., White, R., & Fensham, P. (1988). Developments in style and purpose of research on the learning of science. *Journal of Research in Science Teaching, 25*, 513–530.

Hall, A. R., & Hall, M. B. (1964/1988). *A brief history of science*. Reprinted. Ames, IA: Iowa State University Press.

Hanson, N. (1958). *Patterns of discovery*. London: Cambridge University Press.

Hodson, D. (1988). Toward a philosophically more valid science curriculum. *Science Education, 72*, 19–40.

———. (1985). Philosophy of science, science and science education. *Studies in Science Education, 12*, 25–57.

Holton, G. (1978). *The scientific imagination: Case studies*. New York: Cambridge University Press.

Jones, H. (1977). The past, present, and future of science education before, during, and after the year of the golden-fleeced MACOS. In G. Hall (Ed.), *Science teacher education: Vantage Point 1976, AETS Yearbook*, 189–213. Columbus, OH: ERIC Clearinghouse Science, Mathematics, and Environmental Education: College of Education, Ohio State University.

Kilborn, B. (1980). World views and science teaching. In H. Munby, G. Orpwood, & T. Russell (Eds.), *Seeing curriculum in a new light. Essays from science education*. Toronto: OISE Press/The Ontario Institute for Studies in Education.

Kitchener, R. (1987). Genetic epistemology, equilibration and the rationality of scientific change. *Studies in the History and Philosophy of Science, 18*, 339–366.

Krupa, M., Selman, R., & Jaquette, D. (1985). The development of science explanations in children and adolescents: A structural approach. In S. Chipman, et al. (Eds.), *Thinking and learning skills: Vol. 2*, 427–455. Hillsdale, NJ: Lawrence Erlbaum Associates.

Kuhn, T. (1984). Professionalization recollected in tranquility. *ISIS, 45*:276, 29–33.

———. (1977). *The essential tension: Selected studies in scientific tradition and change*. Chicago: University of Chicago Press.

———. (1962/1970). *The structure of scientific revolutions*. 2nd ed. Chicago: University of Chicago Press.

Lakatos, I. (1970). Falsification and the methodology of scientific research programs. In I. Lakatos & A. Musgrave (Eds.), *Criticism and the growth of knowledge.* London: Cambridge University Press, pp. 91–196.

Laudan, L. (1984). *Science and values.* Berkeley, CA: University of California Press.

———. (1977). *Progress and its problems: Toward a theory of scientific growth.* Berkeley, CA: University of California Press.

Laudan, R. (1987). *From mineralogy to geology: The foundations of a science, 1650–1830.* Chicago: University of Chicago Press.

Linn, M. (1987). Establishing a research base in science education: Trends and recommendations. *Journal of Research in Science Teaching, 24,* 191–216.

Losee, J. (1980). *A historical introduction to the philosophy of science.* 2nd ed. New York: Oxford University Press.

McKenzie, A. E. E. (1973/1988). *The major achievements of science.* Reprinted. Ames, IA: Iowa State University Press.

Mayor, J. R., & Livermore, A. H. (1969). A process approach to elementary school science. *School Science and Mathematics, 69,* 411–416.

Mayr, E. (1982). *The growth of biological thought.* Cambridge, MA: Harvard University Press.

Montagu, A. (1984). *Science and creationism.* New York: Oxford University Press.

Nagel, E. (1960). *The structure of science: Problems in the logic of scientific explanation.* New York: Harcourt, Brace & World, Inc.

National Academy of Sciences and Engineering. (1982). *Report on a convocation.* Washington, DC: National Academy of Sciences and Engineering.

National Science Foundation. (1975). *National Science Foundation curriculum development and implementation for precollege science education.* Washington, DC: U.S. Government Printing Office. 61-5790.

Nersessian, N. (1989). Conceptual change in science and in science education. *Synthese, 80:*1, 163–183.

———. (Ed.). (1987). *The process of science: Contemporary philosophical approaches to understanding scientific practice.* Dordrecht: Martinus Nijhoff.

Novak, J. (1977). *A theory of education.* Ithaca, NY: Cornell University Press.

Novak, J., & Gowin, R. (1984). *Learning how to learn.* New York: Cambridge University Press.

Osborne, M., & Freyberg, P. (1985). *Learning in science: Implications of children's knowledge.* Auckland, New Zealand: Heinemann.

Osborne, R., & Wittrock, M. (1983). Learning science: A generative process. *Science Education, 67,* 489–508.

Piaget, J. (1970). *Six psychological studies.* New York: Vintage Books.

Posner, G., Strike, K., Hewson, P., & Gertzog, W. (1982). Accommodation of a scientific conception: Toward a theory of conceptual change. *Science Education, 66,* 211–227.

Radner, D., & Radner, M. (1982). *Science and Unreason.* Belmont, CA: Wadsworth Publishing Co.

Reingold, N. (Ed.). (1979). *The sciences in the American context: New perspectives.* Washington, DC: Smithsonian Institution Press.

Resnick, L. (1983). Mathematics and science learning: A new conception. *Science, 220,* 477–478.

Rissland, E. (1985). The structure of knowledge in complex domains. In S. Chipman et al. (Eds.), *Thinking and learning skills.* Vol. 2, 107–126. Hillsdale, NJ: Lawrence Erlbaum Associates.

Root-Bernstein, R. (1988). Setting the stage for discovery. *The Sciences,* May/June, 26–33.

———. (1984). On defining scientific theory: Creationism considered. In A. Montagu, *Science and creationism* (pp. 64–94). New York: Oxford University Press.

Rubba, P. (1977). *The development, field testing, and validation of an instrument to assess secondary school students' understanding of the nature of scientific knowledge.* Unpublished doctoral dissertation, Indiana University, Bloomington.

Schneer, C. (1960/1984). *The evolution of physical science: Major ideas from earliest times to the present.* Lanham, MD: University Press of America.

———. (Ed.). (1969). *Toward a history of geology.* Cambridge, MA: MIT Press.

Schwab, J. (1969). The practical: A language for curriculum. *School Review, 78,* 1–23.

———. (1962). The teaching of science as inquiry. In J. Schwab & P. Brandwein (Eds.), *The teaching of science,* 1–104. Cambridge: Harvard University Press.

Schwab, J., & Brandwein, P. (Eds.). (1962). *The teaching of science.* Cambridge: Harvard University Press.

Science 84 (1984, Nov.). 20 discoveries that shaped our lives. *5:*9, 49–155.

Shapere, D. (1984). *Reason and the search for knowledge: Investigation in the philosophy of science.* Dordrecht, Holland: Reidel Press.

———. (1982). The concept of observation in science and philosophy. *Philosophy of Science, 59,* 485–525.

Shapin, S. (1988). The house of experiments in seventeenth-century England. *ISIS, 79,* 373–404, N 298.

Shuell, T. (1987). Cognitive psychology and conceptual change: Implications for teaching science. *Science Education, 71,* 239–250.

Sullivan, W. (1988, Nov. 1). New theories link asteroids' impact to major changes in earth's history. *New York Times,* p. C1.

Suppe, F. (Ed.). (1977). *The structure of scientific theories.* 2nd ed. Champagne-Urbana: University of Illinois Press.

Takeuchi, H., Uyeda, S., & Kanamori, H. (1970). *Debate about the earth.* San Francisco: W. H. Freeman.

Thackray, A. (1980). History of science. In P. T. Durbin (Ed.), *A guide to the culture of science, technology, and medicine.* New York: Free Press.

Toulmin, S., & Goodfield, J. (1962/1982). *The architecture of matter.* Chicago: University of Chicago Press.

———. (1965). *The discovery of time.* New York: Harper & Row.

U.S. Congress. House. (1955). Subcommittee of the Committee on Appropriations. Independent Offices Appropriations for 1956. Hearings 84th Congress, 1st session. Washington, DC: U.S. Government Printing Office.

Von Daniken, E. (1970). *Chariots of the gods?* New York: Bantam Books.

Vosniadou, S., & Brewer, W. F. (1987). Theories of knowledge restructuring in development. *Review of Educational Research, 57,* 51–67.

Welch, W. (1979). Twenty-years of science curriculum development. In D. Berliner (Ed.), *Review of research education.* Vol. 7, 282–306. Washington, DC: American Educational Research Association.

Welch, W., Klopfer, L., Aikenhead, G., & Robinson, J. (1981). The role of inquiry in science education: Analysis and recommendations. *Science Education, 65:*1, 33–50.

West, L., & Pines, A. (Eds.). (1985). *Cognitive structure and conceptual change.* New York: Academic Press.

Wilson, J. T. (Ed.). (1976). *Readings from Scientific American: Continents adrift and continents aground.* San Francisco: W. H. Freeman.

Wittrock, M. (1986). Students' thought processes. In M. Wittrock (Ed.), *Handbook of research on teaching,* 297–314. New York: Macmillan.

Yager, R. E. (1988). A new focus for school science: S/T/S. *School Science and Mathematics, 88,* 181–190.

Index

Note: Page numbers followed by *n* indicate reference notes and source notes.

About the Author

Richard A. Duschl received his Ph.D. in science education from the University of Maryland–College Park, in 1983. Presently he teaches in the School of Education at the University of Pittsburgh. His research has focused on ways to integrate the history and philosophy of science into science-teacher education programs and science curriculum materials. This volume represents a synthesis of the ideas and practices Dr. Duschl has been developing and using in his science methods, curriculum, and science content classes at the University of Houston and Hunter College of the City University of New York.